Eugenics and the Nature–Nurture
Debate in the Twentieth Century

Eugenics and the Nature–Nurture Debate in the Twentieth Century

Aaron Gillette

palgrave
macmillan

First published in 2007 by
PALGRAVE MACMILLAN™
175 Fifth Avenue, New York, N.Y. 10010 and
Houndmills, Basingstoke, Hampshire, England RG21 6XS
Companies and representatives throughout the world.

PALGRAVE MACMILLAN is the global academic imprint of the Palgrave Macmillan division of St. Martin's Press, LLC and of Palgrave Macmillan Ltd. Macmillan® is a registered trademark in the United States, United Kingdom and other countries. Palgrave is a registered trademark in the European Union and other countries.

ISBN-13: 978–1–4039–8422–7
ISBN-10: 1–4039–8422–0

Library of Congress Cataloging-in-Publication Data

Gillette, Aaron, 1964–
    Eugenics and the nature-nurture debate in the twentieth century / by Aaron Gillette.
        p. cm.
    ISBN 1–4039–8422–0 (alk. paper)
    1. Eugenics. 2. Evolutionary psychology. 3. Social ethics. I. Title.

HQ751.G48 2007
363.9′2—dc22   2007003715

A catalogue record for this book is available from the British Library.

Design by Newgen Imaging Systems (P) Ltd., Chennai, India.

First edition: November 2007

10 9 8 7 6 5 4 3 2 1

Printed in the United States of America.

Transferred to Digital Printing in 2009.

For my mother
Roxane Price

Contents

Acknowledgments   ix

Introduction   1

Part I
The Rebirth of Sociobiology
and Evolutionary Psychology

1 Foundations for a "New" Synthesis   21

2 Recent Studies on Human Sexuality   27

Part II
The Birth of Sociobiology
and Evolutionary Psychology

3 The Animal Nature of Humans   41

4 Earlier Studies on Human Sexuality   61

5 Evolution, Ethics, and Culture   95

Part III
The Death of Sociobiology
and Evolutionary Psychology

6 The Rise of Environmental Behaviorism   107

7 Evolutionary Psychology under Attack   121

8 The Death of Evolutionary Psychology   135

9   Lost in the Wilderness   157

Conclusion   163

List of Abbreviations   169

Notes   171

Bibliography   201

Index   219

Acknowledgments

My sincere thanks go to Elof Carlson, Geoffrey Miller, and Jonathan Spiro for discussions on various aspects of this work. Jeff Jackson, Brian Alnutt, David Weiden, and Michael Gelb read and offered very helpful suggestions on parts of the manuscript.

The staffs of many libraries and archives played a key role in the completion of this work. Especially noteworthy is the assistance of the inter-library loan staff of George Mason University and the University of Maryland College Park. Judith May-Sapko, Special Collections Librarian/ Archivist of the Pickler Memorial Library, Truman State University, cheerfully spent many hours helping me obtain documents. Above all, Shannon Cunningham and the inter-library loan staff of the University of Houston-Downtown were truly remarkable in their dedication to obtaining the materials used in this work. Patrick Kerwin of the Manuscript Division of the Library of Congress, Marianne Kasica of the University of Pittsburgh Archives, and Mott Lin of Clark University Archives provided help at critical moments. Holly Heighes graciously provided invaluable assistance uncovering important information in the Robert Yerkes Papers of Yale University. Funding for this project came from George Mason University Department of History and Art History and the American Heritage Center of the University of Wyoming.

Introduction

The keystone of the new social structure, the pivotal factor of advancing civilization, the guide of the new religion, is biology; for man is an animal, and his characteristics, his requirements, his reactions, can be recorded and studied quite as carefully and precisely as those of any other animal.

Edward M. East, Heredity and Human Affairs, *p. 13*

Man has never understood man, and the worst feature of this failure is that his misunderstandings have been fraught with incalculable evils throughout the course of human history.

Samuel Jackson Holmes, "Darwinian Ethics," p. 119

Convictions are more dangerous enemies of truth than lies.

Friedrich Nietzsche, Humans All too Human, *p. 63*

On February 15, 1978, Edward O. Wilson sat on the stage of an auditorium, waiting to address an audience. Suddenly, a young woman leapt onto the stage, grasped a pitcher of ice water, and poured it onto Wilson's head. Others ran onto the stage and waved anti-Wilson placards, chanting, "Wilson, you're all wet."[1]

Several weeks before, two of Wilson's colleagues, Richard Lewontin and Stephen Jay Gould, wrote an attack against him in the *New York Review of Books*.[2] They claimed that Wilson was an ally of eugenicists, and his theory was a dangerous pseudoscience associated with "the enactment of sterilization laws and restrictive immigration laws by the United States between 1910 and 1930 and also for the eugenics policies which led to the establishment of gas chambers in Nazi Germany."[3]

Wilson was a professor of biology at Harvard University. The meeting at which he was attacked was the annual meeting of the American Association for the Advancement of Science.

This is not how scientists are supposed to act. But such was the case with the nature–nurture debate of the twentieth century.

When the century opened, most biologists were convinced that much of human behavior had a hereditary basis. Over the course of the next three decades, a coherent body of research developed from the union of that premise with the concepts of natural and sexual selection as conceived by Charles Darwin the century before. By 1930, a group of biologists had substantial evidence to support the premise that some aspects of human behavior were the products of evolutionary adaptation. By then, the theory of evolution in one form or another was almost universally accepted in the scientific community. And humans were, after all, animals. It seemed that some of the mysteries of human behavior were finally yielding to the insights biology could provide. Biology helps us understand ourselves, in body and in mind. How could it be otherwise?

Contrary to what we might wish to believe, science does not always follow clear paths illuminated by reason. Certainly the nature–nurture debate of the early twentieth century stands out as an example of science guided by ideological preconceptions. Rivalries, hatreds, and insinuations poisoned the science of human behavior in the years after World War I.

In the first thirty years of the new century, many scientists studying human behavior were amassing evidence that some aspects of human behavior were influenced by heredity, driven by evolutionary forces. Charles Darwin had suggested that such was the case. Experimental evidence and theoretical refinements over the next several generations strengthened the argument. However, a strange twist of fate emerged to damn the work of these evolutionary psychologists. They were, with few exceptions, also eugenicists. Their work had been inspired by their belief in the promises of eugenics to mold the human species.

Eugenics was a pseudoscience, an epiphenomenon of a number of sciences, which all intersected at the claim that it was possible to consciously guide human evolution. Springing from the platform that human behavior was tied to evolutionary heredity, eugenics made the disturbing claim that human traits could be accurately measured, quantified, and assessed for social desirability. Given that these traits were supposedly inherited, manipulating the reproduction of people with these traits could spread or diminish the prevalence of these traits in later generations. As became increasingly clear, eugenics could be dressed up as an apparently humane solution to a variety of social and medical problems. However, careful consideration of its methods and aims revealed deep layers of flawed assumptions and immoral agendas. Some scientists refused to see the moral dangers inherent in eugenics. To them eugenics seemed to offer the promise of a world engineered by coolly logical scientists. Such a world, they naively assumed, could only be for the better. Thus, with one hand they offered the world their scientific work as evolutionary psychologists.

With the other they blithely offered a eugenical program to make practical use of the new knowledge and remold the human species. It may have been possible to evaluate and accept the scientific work, while sternly rejecting the pernicious eugenic fantasies. This did not happen.

There were other students of human behavior in the early twentieth century who had never accepted that "nature" had much to do with the human mind. This group was wedded to the assumption that environment shaped virtually all behavior. They caught on quickly to the deep flaws of eugenics, and did the world a service by exposing these flaws. However, their distrust of the possible links between heredity and behavior, though in some ways salutary, became exaggerated. Environmental behaviorists[4] eventually attacked any suggestion that human behavior could be influenced by heredity. Their "anti-nature" attitudes, driven by ideological conviction, stifled important early work on evolutionary psychology. By the 1930s, the nature–nurture debate was definitively resolved in the United States and Britain. "Nature" theories of human behavior were now seen not simply as bad science, but as *immoral* science. Human nature was only—nurture. Those who questioned this new truth did so at their own professional peril.

By the 1960s, however, nagging questions demanded answers. Since World War II, a growing body of evidence supported the existence of instincts in animals. Did this not have some application to human beings? Had not scientists once agreed that humans were, in the fact, animals?

In 1975, Edward O. Wilson published *Sociobiology: The New Synthesis:* This work was an extensive compendium of recent research that gave the world a new "nature" theory called sociobiology. Its sister theory, evolutionary psychology, would be born a decade later.[5] And so the nature–nurture battle once again raged in the scientific world.

However, the casualties from years before would not be resurrected. The early pioneers in the biology of human behavior, whose solid scientific work had been tainted by their simultaneous advocacy of eugenics, remained forgotten, their work unknown. In a sense, sociobiology and evolutionary psychology had died before World War II, only to be reborn decades later in a new guise. Unbeknownst to the new scientific generation, some of the most important theoretical and empirical work conducted in the late twentieth century in the name of sociobiology and evolutionary psychology had already been accomplished half a century earlier.

This book will examine the vicissitudes of the nature–nurture debate of the twentieth century, and the ideological underpinnings of the debate. On the broader scale, it will consider the extent to which science, in its "real world" context, was driven by ideological imperatives in the twentieth century.

Many in the American public retain the notion of the "scientist" as an individual whose life is devoted to the discovery and advancement of knowledge, perhaps oblivious to the ethical implications of his or her discoveries. Scientists, supposedly, are unafraid to challenge the boundaries of social convention and the current mental frameworks that support those conventions. Americans are taught in high school that science is a logical system for testing assumptions about nature, without regard for the biases, wishes, or desires of its practitioners. Both the scientific method and the scientists who practice it are supposedly paragons of objectivity.

Another basic element of American culture is the belief that science must always "advance" some area of knowledge and yield useful applications in technology. To Americans, it is axiomatic that science will make the future faster, smaller, healthier, and more wondrous. Nature's mysteries will continually vanish before us, and undreamt of discoveries come ever closer into view.

This book will argue quite the opposite. As Catholic theologians say about their Church, "though the institution is perfect, the individuals who compose it are not."[6] I would echo those words. The scientific process, at its purest, is unsurpassed in its ability to create useful knowledge; but scientists, as individuals, all too often are incapable of attaining such a goal. Thus, a fundamental tension exists between the creators and the process of creation.

While considering the history of the nature–nurture debate, we will constantly face the question of objectivity in science. Is it possible for a scientist to be objective, to separate himself or herself from the viewpoints and constraints of his or her cultural or social group when attempting to interpret the natural world? Indeed, is objectivity a goal of science at all? How can we know when a scientist is or is not objective?

Philosophers of science have long been concerned with the issue of scientific objectivity. The debate, in its simplest form, tends to resolve itself down to two theories of scientific knowledge. Scientific realists believe that scientists are capable of objectivity, that science progresses largely through the rational course of experimentation and debate intrinsic to the scientific method, and that science discovers the true reality of nature.

On the opposite side of the spectrum, scientific antirealists focus on the ability of science to solve problems, rather than telling us anything about the nature of reality. There are a wide variety of antirealist positions. Neo-Kantians argue that our understanding of nature is dependent on the structure of the human mind. The mind structures reality in a way that "makes sense" to humans, but does not necessarily reflect an external reality.

Paradigm-relativists believe that a scientist applies a theory to interpret data not because of the merits of the theory in explaining the data (all too often competing theories offer equally satisfying explanations) but on the basis of his or her own worldview or paradigm. Science is not a divine mystery found among the crystal spheres; it is an intellectual process that cannot entirely free itself from the society in which it operates. As such, the research questions chosen for study, the hypotheses thought up to answer these questions, the methods employed to obtain the data, the interpretation of the data, and the conclusions drawn from this interpretation will tend to reflect the social, cultural, and intellectual currents present in that society. We might say that science cannot be entirely independent of the *volksgeist*. Even professional loyalties, friendships, and the desire for career advancement can sway the course of scientific investigation away from an "objective" view of nature, if such exists.[7] Of course, the scientist's own ethnicity, culture, or gender may be an influential factor. In this view, science is not an expression of rationality and objectivity. It is nonrational or even irrational.

Pragmatism derives from the pragmatic philosophy popularized at the turn of the twentieth century by William James and John Dewey. Here, it is conceded that certainty is unattainable. All we can hope to achieve is an approximation of the truth. Pragmatists are concerned with the ability of a theory to predict outcomes. If it is successful at this task, it is regarded as "approximately true." In effect, the more we can successfully manipulate nature, *do* things because of a theory, the more likely that a theory is approximately true. Resolving debates about the ultimate truth of reality is not an objective in pragmatism.

Regardless of the version of scientific antirealism selected, all tend to suggest that objectivity in science is an elusive goal (if it even is a goal). Because of the structure of the human mind, the social or personal reasons for embracing one theory instead of another, or the concentration on the practical applications of science, scientists are not objective. Their understanding of reality does not necessarily correspond to the real.[8]

One undeniable product of science in the modern world is the increased power it has given humankind over nature. By the twentieth century, the ability of science to exercise control (over nature, society, or the individual) was truly awe inspiring. It was also frightening, as the development of poisonous gases in World War I and the atomic bomb in World War II would demonstrate. As the financial and resource costs associated with scientific investigation soared along with the potentially beneficial or destructive results of scientific discovery, democratic societies increasingly debated the issue of control *over* science. How could science be controlled so as to yield the most socially beneficial results? Should individual scientists be entirely free to follow their own inclinations and investigate whatever questions they chose to ask of nature? Should some unit of government, body of citizens, scientific board, or combination of such entities govern the production of science? What entity competent to regulate science best embodied the values of society? One inherent difficulty in deciding who should regulate science is determining who best represents the common social values that should be applied to science. In a culturally diverse society, what appears as "common social values" to one group might be interpreted as a particularistic view of the world to another group—again, the question of bias surfaces.

The contest between the social control of science and the freedom of scientists to pursue their own research objectives has ancient roots. In 415 CE, the citizens of Alexandria, Egypt, driven by Christian religious fanatics, massacred the neo-Platonic scholar Hypatia because they felt threatened by her political and popular influence. Christian authorities of course continued to exercise a severe degree of control over scientific curiosity. One only need think of the cases of Giordano Bruno and Galileo Galilei. Both men were punished for their insistence in privileging their rational understanding of the universe above Church dogma.

The later shift in power to the state in most of the world did not necessarily alleviate the pressure on scientists to conform their ideas to the prevalent ideology. Perhaps the notorious Lysenko Affair is the best example of state interference in scientific investigation. Trofim Lysenko, a Soviet biologist of peasant origins, claimed that he had discovered various means to cause remarkable improvements in agriculture based on a confusing mixture of ideas borrowed from Lamarckianism (more on this later), Darwinianism, and hybridization. His promise of cheap, practical advice to peasants to improve yields was endorsed by Stalin in the 1930s. Lysenko used his power to hound his critics and destroy Soviet genetics for a generation.

Nothing in American history approaches the chaos of the Lysenko affair; however, we might note that recent issues of genetic engineering and the definition of "human life" from an ostensibly biological viewpoint have elicited an enormous degree of intervention by all branches of the American government at federal, state, and local levels.

In the United States, organized control over science has been exercised as often by scientific bodies as by the state *per se*. This book will elucidate the fierce clash between loosely organized, ideologically driven scientific groups over the nature–nurture debate. I would assert that the successful drive to squelch dissent by those who believed that virtually all human behavior was environmentally determined was not only detrimental to the mission of science, which is to acquire a working understanding of nature, but was implicitly mistrustful of democratic society. I believe that a democratic society should accord to scientists a high degree of freedom to pursue research. Certainly, oversight bodies should exist (as they do) to insure that the potential importance of knowledge to be gained through such research is in accord with the amount of resources to be utilized in conducting the investigation, that near-universal norms of human and animal rights are protected, and that potential adverse consequences of the research are minimized. Moreover, I believe that a much higher level of oversight should be exercised by the community when it is a question of the *uses* to which this knowledge will be permitted. This distinction may seem overly reductionist—the dividing line between "pure" science and technology is not so obvious. This book does not seek to offer specific guidelines for the socially beneficial conduct of science. It does strive to demonstrate that in at least one case scientific advance has been stunted because of ideological conviction. If one wishes to draw a prescriptive lesson from this episode, it would perhaps be a plea for democratic societies to be brave enough to allow scientists to seek knowledge while operating under the minimum restraints necessary to insure that nearly universally held norms are not violated. However, the application of such knowledge in creating policies, programs, therapies, laws, and regulations should be stringently controlled by representatives of the diverse groups who have an interest in the utilization of this knowledge.

There is perhaps no better exposition of the tensions inherent in controlling science than in the nature–nurture debate of the twentieth century. In this on-going debate, science confronts some of the most basic human questions: What are human beings? Why am I as I am? From where do my desires, behaviors, and goals originate? When considered in the aggregate, how do such motivations and behaviors shape society? Can we overcome those behaviors that we deplore, such as violence or cruelty? How useful is the comparison between human behavior and the behavior of the higher primates? Do apparently similar behavioral patterns found in many cultures imply an underlying biological cause? Likewise, do frequently noted differences between the sexual behavior of many men and women suggest gender-stereotyped socialization or sex determined mating behaviors largely dictated by biology ("sexual essentialism")? These are the questions that engender passionate convictions, born of the quest of self-identity. The nature–nurture debate proved incapable of seeking answers to these questions without invoking deep-seated beliefs that carried ready-made answers. In this case, many brought to the data a predetermined faith in the dominance of biology or environment in guiding human behavior. So great was the strain of preconceptions on the process of scientific discovery that research in this area nearly ground to a halt for some forty years. As will become apparent, rigid opposition to the notion that *any* meaningful human behaviors could be inherited stifled many potentially fruitful avenues of research and thus inhibited the progress of psychology and anthropology in the mid-twentieth century.

Bitter controversy has always surrounded the nature–nurture debate. In the early nineteenth century, liberals tended to accept John Locke's view of the human mind as a *tabula rasa*, or blank slate.[9] Personality and morality were products of the individual's interaction with their social environment. More conservative thinkers often stressed the innate qualities of character. As scientific racism developed in the late nineteenth century, this belief in innate character easily accommodated itself to the frequently ludicrous racial stereotypes assigned to various ethnic groups.

Charles Darwin's theory of evolution, first announced in 1859, certainly did not lessen the divisiveness. Darwin offered strong evidence to support his claim that individuals in every species tended to compete for limited resources in a harsh world. Any organism that had hereditarily transmissible traits which allowed it to survive and reproduce more successfully than its competitors would perpetuate these traits into succeeding generations through their descendants. Over time, the most successful combinations of inheritable traits would spread throughout the population. The long-term effects of these changes, multiplying in number and affect so long as they increased survivability and reproductive success, would eventually create organisms substantially different from their remote ancestors. Thus, if two original populations of the same species were separated in some manner, and exposed to different environmental pressures, two distinct species might eventually evolve, given a very long period of time. This process came to be known as "natural selection."

Darwin found the traits of most organisms explicable through this process. But not all. For example, the exuberant tail of the male peacock could only be seen as a physical hindrance and a drain of energy resources. Elk, birds, and other creatures also sported apparently nonessential but highly flashy structures. So some other mechanism must account for such unusual physical features, Darwin reasoned. Noticing that such features often seemed to be characteristic of males, and attractive to females, Darwin hit upon another evolutionary mechanism: sexual selection. Here, it seemed reasonable to suppose that any physical structure connected with overall health and reproductive fitness, if perceived as attractive by the opposite sex, would likely enhance the reproductive success of the attracted partner. If both sexes responded to such markers, their reproductive success would be greater still. And since reproduction is the *sine qua non* of spreading particular inheritable traits through a population over time, this process of sexual selection should act much like natural selection: traits that better adapted a species for survival and reproduction in its natural environment would likely increase in frequency over time, thereby permanently changing (or "evolving") the species.

So powerful were Darwin's evolutionary theories, and so useful were they in understanding the natural world, that they equal if not surpass in importance Copernicus's model of the solar system or Newton's theory of gravity. However, since they also addressed the causes of human behavior in a much more intimate fashion than any other major scientific theory before them, they also generated that much more controversy. In general, the idea of evolution was so controversial because it was at the same time so incredibly successful in explaining a vast array of biological phenomena, while also being entirely blind to supernatural forces or ethical concerns. In short, evolution simply had no place for moral or religious agendas.

The most serious problem with Darwin's theories of evolution was that he could neither demonstrate a convincing mechanism by which traits could be passed on from one generation to another, nor articulate any laws that governed this inheritance. Other scientists, while fully agreeing that evolution of some type characterized all life, conjured up other possible mechanisms of change besides natural and sexual selection. Even the rediscovery and rapid acceptance of the Mendelian laws of inheritance (what we now call "genetics"), after 1900 failed to quash non-Darwinian evolutionary theories. It was only with the experimental failure of other evolutionary models that the so-called Great Synthesis of evolutionary theory occurred, around 1940. The Great Synthesis combined the Darwinian theory of natural selection with Mendelian genetics, and mathematically demonstrated that both processes were in full accord with one another. Thereafter, the Great Synthesis, based on Darwinian natural selection, became universally accepted among biologists as the one and only believable theory of evolution.

One reason that Darwinian natural selection was resisted for so long was that it carried so many ideological implications. For one, it seemed to suggest that bitter competition and life-draining struggle were natural. If psychological traits were as inheritable as were physical traits, then Darwin's theory seemed to suggest that self-conscious efforts at human mental and ethical betterment would not last beyond a person's lifetime. Socialists, who welcomed Darwinian evolution's lack of a role for a deity, nevertheless found its embrace of individual or group struggle very distasteful. If there was a hierarchy of abilities in species and between species, it seemed but a small step to allege that modern human society also had a "natural" hierarchy. Darwinian natural selection seemed to predict a future of perpetual human struggle.[10]

Darwin's concept of sexual selection also faced a cool reception. Several cultural forces arrayed against it, for a variety of reasons. In its earlier years, there seemed to be a reticence to accord sex such a strong motivational force in evolution. Some biologists disparaged sexual selection as a minor variant of natural selection, which they saw as the chief driving force of evolution. Thus, they tended to blindly overlook the evidence for the operation of sexual selection that existed throughout nature. It was undeniable that some species had distinctive sexual markers. But many biologists interpreted these markers as a visible means to intimidate other males without the necessity for physical combat. Others thought that sexual ornamentation was meant to identify members of the opposite sex by others of the same species, in order to avoid wasteful non-procreative sexual contact. The concept of "group selection," the possible existence of which was even lukewarmly endorsed by Darwin, argued that a population whose members aided each other while opposing those of other species might be most likely to survive.[11] Furthermore, sophisticated statistical analysis had to be applied to observations of suspected sexual selection to demonstrate its existence.[12] Later, liberal-minded scientists found the sexual stereotyping implicit in theories of human sexual selection distasteful. The growing conviction in the scholarly community that human behavior was dictated by social and cultural forces, rather than by biology, reinforced this view.[13] However, as this book will show, there was a greater interest in conducting sophisticated studies on human sexual selection in the decades before World War II than historians have hitherto suspected.

For those who were inclined to accept the idea of natural human hierarchies or who saw personality traits as inheritable, Darwin's theory was something of a godsend. Here was serious scientific justification for their beliefs. Many scientists from both these groups also embraced eugenics as a way to accelerate the mechanisms of Darwinian evolution to "improve" the human race.

Eugenics was a prescription for guiding human evolution authored by Charles Darwin's cousin, Francis Galton. Galton, whose principle goal in life was to render everything mathematically describable, hit upon the idea that tests of various sorts could be devised to rank individuals for intelligence, ambitiousness, industriousness, and the like. If individuals who scored high in those traits desired by society could be persuaded to mate with one another, and prolifically bear children, and those who fell below socially desirable standards could be dissuaded from mating, then human evolution could be artificially accelerated while guided by wise scientists in whatever direction was deemed desirable.

Eugenics appealed to scientists with almost any political ideology. To become a eugenicist, one had to believe that application of knowledge of evolution and genetics to social engineering could ultimately improve human life and accelerate social progress.[14] However, it would be disingenuous to imply that eugenics did not have a particular appeal to any specific ideological group. Most often, eugenicists believed that human personality was formed more by heredity than by environment. There were also many racists among the eugenicists who sought to vigorously defend their racism through appeals to science.[15]

Scholars today debate the degree to which eugenics in the United States and elsewhere was inextricably linked to the horrors of Nazi race hygiene. Stefan Kühl and Allan Chase, in particular, do not see much meaningful difference between the assorted manifestations of eugenics in various countries in the decades before World War II and the theoretical and applied eugenic policies of Nazi Germany. Others take a more nuanced view. They recognize that there was a wide variety of eugenic theories, some of which were much less race- or class-based than others. Eugenicists might also give greater or lesser acknowledgment to the role that environment played in shaping human behavior. In some cases, eugenics was almost imperceptibly intertwined with health care, child care, birth control, and sex education issues. In this sense, eugenics has been called, "a 'modern' way of talking about social problems in biologizing terms."[16] *Eugenics and the Nature–Nurture Debate in the Twentieth Century* reinforces the observation that eugenics infused itself into much of the biomedical and evolutionary discourse of the early twentieth century.

Eugenicists were nothing if not organization builders. A number of overlapping eugenic organizations emerged in the United States in the first thirty years of the twentieth century. The Galton Society was foremost among them. The Society originated in 1919, and met a number of times each year at the American Museum of Natural History. There the members listened to lectures by leading eugenicists or evolutionary psychologists, and debated the major eugenics issues of the day. This inevitably included racial science and immigration restriction. The Galton Society also served as an informal coordinating committee for all organizations with substantial eugenic interests.

With few exceptions, the roster of the Galton Society reads like a "Who's Who" of leading American academics. The Society was founded by the wealthy environmental activist and virulent racist Madison Grant; Charles Davenport (the director of the Eugenics Record Office of the Department of Genetics at the Carnegie Institute of Washington); and Henry Pratt Fairchild (the director of the American Museum of Natural History). These founders then admitted, as charter members, six of their colleagues in the academic community: biologist Edwin G. Conklin of Princeton University; anatomist George S. Huntington of the College of Physicians and Surgeons; zoologist J. Howard McGregor of Columbia University; Edward L. Thorndike, a leading psychologist at Columbia University; paleontologist William K. Gregory of the American Museum of Natural

History; and another paleontologist, John C. Merriam, of the University of California (Berkeley). Later on, other members were inducted into the Society. Most were renowned professors at Harvard, Yale, Princeton, Columbia, or the American Museum of Natural History.[17] These men or academics of similar stature served at one time or another as presidents of the American Eugenics Society, which was the principal eugenics organization for the larger academic community.

American eugenics societies reached their full flowering at the same time that their ideologies were threatened by a rising force in the social sciences: environmental behaviorism. Known simply as "behaviorism" in psychology and sometimes called "Boasianism" by cultural anthropologists, environmental behaviorism took as its basic premise the idea that human beings had evolved mental faculties so far in advance of other animals that the basic laws of evolution no longer applied when it came to human behavior. Humans, in a sense, were entirely self-created; they were not subject to instincts, as other animals might be. Also, an increasing number of anthropologists saw a wide variation of cultural behavior in different human societies, with no apparent underlying foundation to explain them. This suggested that human behavior did not rest upon a universal biological stratum.[18]

This view, incidentally, also liberated the social sciences from any association or reliance on the biological sciences within academic institutions. Only the social sciences, with their unique perspective on humanity, could truly interpret human behavior on an individual or social level. Therefore, the social sciences were worthy of a great degree of professional status and financial support.[19]

Environmentalists were fundamentally egalitarians. Therefore, they wisely rejected the malevolent racial science and simplistic hereditarian hierarchies of some evolutionary psychologists. They also were very skeptical that the intelligence tests then in use measured much more than the degree to which an individual was educated by mainstream standards (in this they were right). Environmentalists also suspected that notions of "psychobiology" appealed to nationalistic chauvinism, class elitism, racial superiority, and anti-Americanism. Above all, America stood for democratic equality. After 1933, when the Democrats swept into power on the basis of Franklin Roosevelt's New Deal, the voices of those committed to advancing liberalism in America grew louder. Many evolutionary psychologists, on the other hand, emphasized the idea that there are variations in individual personality and capabilities that are inherited and can vary only so much through changes to the social environment. Such ideas appeared reactionary; they were not congenial to the new political and cultural mindset.[20]

This egalitarian mission seemed all the more important given the events in Europe in 1933. Germany was moving in the opposite political direction. Adolf Hitler and the Nazi party assumed control over Germany in January of that year, and rapidly constructed a totalitarian dictatorship. As Rudolph Hess would claim, Nazism was "applied racial science."[21] The Nazi propaganda machine was fond of proclaiming that "Biology Is Destiny." Their version of biological truth was embedded in racial "science" and eugenics. The Nazi regime implemented eugenic sterilization and racial apartheid so swiftly and thoroughly that it left Germany's American admirers in the eugenic movement almost breathless. However, the fanaticism and brutality accompanying Hitler's policies could not be entirely hidden from the world's press. The fundamental ideological clash with the United States was obvious. To American scientists, the political toadying of Germany's scientists at international conferences was particularly distasteful. The more Germany embraced eugenics in the 1930s, the more the movement gained an unsavory reputation in the United States.[22]

With all of these forces arrayed against it, evolutionary psychology, by the end of the 1930s, was doomed. Its last supporters were concentrated in the Axis countries, and often espoused twisted, racist versions of evolutionary psychology. With their defeat in 1945, evolutionary psychology seemed to belong to an entirely discredited past. Another thirty years would pass before evolutionary psychology was again considered based on its own merits and defects rather than on the ideological and political implications that had surrounded it before World War II.

The reader will of course notice that the thirty-year battle between these scientific theories regarding human behavior had very little to do with science, and very much to do with ideological and political conflict. Through the rest of the century, this dynamic slowly loosened its grip over the behavioral sciences, though never entirely cease influencing the debate.

By the 1970s evolutionary psychology began to revive. There are two important reasons for this. For one, the generation of scientists active in the interwar years had almost all retired or died. The fear that evolutionary psychology was eugenics in a different guise had diminished. The extraordinarily limited amount of professional history taught to scientists in the universities aided the intensification of this collective amnesia regarding the uncomfortable history of eugenics and racial science that had tarnished biology. Younger biologists of the 1970s and later might not even *know* what eugenics was. There was now sufficient calm in the scientific community to allow at least the new science of sociobiology a decent hearing. Edward O. Wilson provided the foundational text. It was a wise move on the part of Wilson to stress the behavior of nonhuman animals in his first book on sociobiology. This gave its concepts more of an opportunity to be considered and allowed the reader to draw the obvious analogies to human behavior. Ironically, a similar tactic had to be taken by Charles Darwin in *his* first book on evolution 117 years earlier, to placate a potentially hostile public reaction to his work.

More importantly, sociobiology and evolutionary psychology revived in the 1970s simply because the accumulated evidence of the existence of instinct, and brilliant theories relating genetics to behavior, could no longer be conscientiously ignored. Chapter 1 of this book will demonstrate that ethologists (animal behaviorists) and zoologists were embracing sociobiology long before Wilson ever published *Sociobiology*. The major concepts were in place by the early 1970s for the new sciences of sociobiology and even evolutionary psychology. It merely took the efforts of those brave and determined enough to announce their birth, and study human behavior from their perspective, to catapult sociobiology and evolutionary psychology once again into the center of the debate regarding the causes of human behavior.

Of course, there were still some who reacted violently to any resuscitation of evolutionary psychological theories, as we saw above. The evolutionary psychologist David Buss relates one of many instances of the intense clash between those who were willing to accept the thesis that violence and conflict seem to be endemic in human society and those who refused to accept such a conclusion:

> When the anthropologist Napoleon Chagnon [and James Neel, in the 1970s] documented that 25 percent of all Yanomamö Indian men die violent deaths at the hands of other Yanomamö men [and receive more wives as a result], his work was bitterly denounced by those who had presumed the group to live in harmony. The antinaturalistic fallacy occurs when we see ourselves through the lens of utopian visions of what we want people to be. . . . Despite the evidence, people cling to these illusions.[23]

Terrence Turner and Leslie Sponsel described Neel and Chagnon's work with the Yanomamö as, among other accusations, a product of Neel's belief in "fascistic eugenics . . ."[24]

So fierce might the conviction in fundamental human innocence be held that the linguist Steven Pinker likens it to the zeal ordinarily reserved for religion: "Just as religions contain a theory of human nature, so theories of human nature take on some of the functions of religion, and the Blank Slate has become the secular religion of modern intellectual life. . . . Challenges to the doctrine from skeptics and scientists have led others to mount the kinds of bitter attacks ordinarily aimed at heretics and infidels."[25]

Some social scientists, especially those old enough to remember the eugenics of the pre–World War II years, feared that the new sociobiology was a revival of the hereditarian doctrines that served as the basis of eugenics and racial science. They worried that the new discipline might be used to justify racism, the genetic superiority of elites, the inferiority of Third World immigrants, or any number of evils that they had hoped were finally disappearing from Western society. The seriousness of their charges certainly warranted a careful examination of sociobiology to insure that it did not embrace such nefarious goals.[26] As we shall see, such fears proved groundless.

In light of the concerns of their critics, it is important to reconsider what sociobiology and evolutionary psychology *are* and what they *are not*. They are most emphatically *not* a return to the discredited theories of racial science and eugenics. These scientifically inspired "movements" were fallacious enough to be discredited on their own lack of scientific merit, without even considering their distressing ideological implications.

Evolutionary psychology also does not deny that environment plays a very important role in guiding the development of human behavior. Indeed, environmental adaptation is considered the foremost driving force in evolution. However, evolution works only through very slow changes and adaptations. Human culture has created civilization in a much shorter time period than the process of evolution can create biological adaptations to new conditions. Thus, to some extent, human beings are primitive peoples born into a highly developed civilization. In such a case, biological adaptation may be very much behind the times, and incapable of successfully adapting to the incredibly rapid changes produced through technological and social development. However, thanks to the great degree of "built-in" adaptability that human beings have evolved (in other words, human beings can easily "learn"), they are still capable of functioning relatively effectively in their new environment.[27]

Sociobiology and evolutionary psychology see human behavior as the result of the interaction of inherited behavioral tendencies with the current environment in which a person finds himself or herself. Neither genes nor culture can fully explain the whole range of human behavior. However, genetic analysis and behavioral studies based on evolutionary theory can more fully understand to what extent behavioral tendencies are conditioned by heredity. These sciences also are interested in exploring the possibility that cultural universals (aspects of culture that seem to be shared by all societies) are influenced by evolved behavior patterns.[28]

Evolutionary psychology sees human beings as animals, just like any other animal. The human species has evolved both physically and psychologically over time. Thus, sociobiology and evolutionary psychology are interested in examining human beings as a *species*; they have no interest in looking at individual behavior, and have no reason to suspect that there are any collective group behavioral differences, as would racists. Sociobiology and evolutionary psychology have absolutely no interest in creating ideological movements or becoming involved in politics. With a few notable exceptions, they even eschew making policy recommendations to improve society.[29] Such is the beneficial result of the reaction against eugenics and racism.

Rather, sociobiology and evolutionary psychology seek to restore human beings to their place in the animal world. Specifically, they seek to understand the manner in which natural processes, such as evolution, have affected the evolution of human behavior. In this endeavor, they have restored Darwin's concept of sexual selection to an important role in evolutionary theory. Indeed, much of the astounding work in the past two decades in evolutionary psychology has resulted from studying the implications of sexual selection on human behavior.[30]

However, evolutionary psychology was not always in such benign hands. Given the association of evolutionary psychology and eugenics a century ago, one is naturally left to ask if untainted scientific knowledge can be produced by morally objectionable scientists. Can "good" science be produced by morally corrupt scientists? In his book *The Nazi War on Cancer*, Robert Proctor acknowledged the moral minefield that surrounds "good" science produced by "bad" scientists: in this case, the research on cancer and other health issues by pro-Nazi scientists. In the end, Proctor can only offer an ambiguous answer to the question of the relationship of science to the scientist: "Perhaps what is needed is a severing of the already frayed ties that once were said to conjoin technical and moral virtues; I'm not really sure."[31]

Growing out of his interest in the cultural mediation of scientific truth, Proctor has created a new subdiscipline of historical study, which he calls "agnatology." Agnotology he defines as "structural apathies, communities of disinterest, and the social production of ignorance."[32] As an example of agnotology, Proctor offers an example of the research conducted under Nazi auspices in the 1930s and 1940s showing that smoking was a major cause of cancer. However, the postwar condemnation of the horrors of Nazi eugenics and human experimentation engulfed even the useful research on lung cancer.[33] So great was the revulsion associated with the Nazi regime that all German science produced during the Nazi period, even those few results that might have had some legitimate social utility, were wholly discarded.

Although the United States has never experienced such a wholesale rejection of its past as has Germany, it can nevertheless offer other examples of "agnotology." In particular, the 1920s witnessed the conservative moralistic condemnation of alcohol and tobacco use. Scientists buttressed moral rejection of these "recreational drugs" with studies showing their medically harmful effects. However, by the 1930s the growing awareness that the attack against alcohol and tobacco had a strongly conservative foundation led more liberally inclined scientists to react against earlier research purportedly recognizing the health hazards of these substances.[34] As Proctor explained, "it was rare to find an American physician who criticized tobacco in the 1930s or 1940s, and those who did were often dismissed as prudes or cranks."[35] A new generation of scientists, approaching these problems without the benefit of earlier research, began in the 1950s to suspect that cigarette smoking might indeed contribute to lung cancer.[36] In the 1960s and 1970s, medical researchers rediscovered that excessive alcohol consumption could contribute to cirrhosis of the liver, cardiomyopathy, adverse fetal effects, and esophageal cancer. This had all been known in the early twentieth century, yet had been forgotten for cultural and ideological reasons.[37] This book will offer another equally profound example of "agnotology," as it relates to the overthrow of genetic theories of human behavior in favor of learning, culture, and other environmental causes of behavior. Once again, the shift in belief will not emerge from scientific discovery, but from ideological change. And like the studies above, the consequence will be the deliberate forgetting of useful knowledge.

Before we begin our exploration of the nature–nurture debate, an essential question posits itself. If ideology has been such a dominate a force in the nature–nurture debate of the last 150 years, how can we know that the relatively strong position of evolutionary psychology today is not due simply to a temporary ideological ascendancy? Through all the vicissitudes of the struggle to understand human behavior are we getting any closer to the truth? The difficulty in answering this question is all the more serious when we consider the similarities between the advocates of nature and those of nurture. The debate was structured by the modern academic environment. In the early twentieth century both evolutionary psychologists and environmental behaviorists were inspired by their own ideologies to investigate human behavior from particular perspectives. Both groups amassed evidence and developed theoretical explanations for their observations. Not coincidentally, these theories were influenced by their favored ideology. Individuals from both groups sought to undermine the careers of their opponents using sometimes distasteful means. Nevertheless, both groups operated in the context of scientific institutional oversight, peer review, and public discourse.

In the early twenty-first century, the current incarnations of evolutionary psychologists and environmental behaviorists still fiercely contend the role of nature and nurture in shaping human behavior. However, participants in the debate today are, it seems likely, better grounded in a common moral framework than their predecessors. Numerous scientists and scholars on both sides of the debate are well-meaning, seek to determine the truth as closely as possible, and have amassed very impressive evidence to support their cause. On what basis can we logically choose between the versions of reality that each side offers?

Perhaps we can agree that science is an enterprise that seeks to interpret nature in such a way as to render it predictable and useful. This definition, at the heart of pragmatism, can aid us in determining the relative merits of evolutionary psychology or environmental behaviorism at any one time. Over time, it became increasingly apparent that some claims of evolutionary psychology simply predicted and explained human behavior better than could environmental behaviorism. One of the main reasons behaviorism ultimately collapsed was because its conclusions simply did not correspond to the mounting evidence in favor of "human nature" as a powerful force influencing some aspects of human behavior. Behaviorism could neither predict nor explain the results of numerous experiments on the higher mammals and humans through an appeal to environmental conditioning alone. This engendered a paradigm shift in favor of the existence of human instincts in conjunction with environmental influences. In other words, the "nature" theory of evolutionary psychology currently explains the observed facts better than does the "nurture" theory of behaviorism for some human behaviors. In the end, scientific logic has prevailed over ideology, but only after decades of delay and the loss of important scientific work.

The search for the proximate truth about human behavior is far from over. Certain subfields of evolutionary psychology (e.g., human sexual behavior) have a stronger scientific basis than do others (e.g., explanations for Islamic suicide bombers).[38] The critics of evolutionary psychology, especially feminists, should be encouraged, and their arguments given the greatest consideration.[39] That the continuing search for knowledge in the nature–nurture debate might eventually turn against evolutionary psychology is possible. However, we can hope that the greater sensitivity to the role ideology plays in science, through the work of Robert Proctor and others, will prevent scientific casualties such as those suffered seventy years ago.

Chapter 1 will look at the progress sociobiology and evolutionary psychology have made in the past thirty years in helping us understand human behavior. Thereafter, this book will concentrate on revealing how much of this knowledge had once been known—then lost.

Part I

The Rebirth of Sociobiology and Evolutionary Psychology

1

Foundations for a "New" Synthesis

Conventional histories of sociobiology and evolutionary psychology, such as those written by Edward O. Wilson, credit a group of three European ethologists (animal behaviorists) as the "fathers" of these sciences. All three began their work in the 1930s, and continued researching and writing into the 1960s.

Niko Tinbergen was a Dutch-born ethologist who accomplished much of his important work at Oxford University, where he obtained a position in 1949. Working with stickleback fish and with gulls, Tinbergen eventually decided that behavior was as important as physical traits in understanding the evolutionary relatedness of different species. Two species whose behaviors were similar might well have a close evolutionary relationship. In 1951, Tinbergen published one of his most important works, *Study of Instinct*, which argued that much of animal behavior was instinctive.[1]

The Austrian ethologist Konrad Lorenz also worked with birds, occasionally collaborating with Tinbergen. Lorenz developed the concept of "imprinting": during a critical period in their lives, the young of some animal species will firmly fix their parents' physical identity in their minds. This information will lead them to follow their parents early in life, and later will play a role in the identification of potential mates, flock formation, and participation in other social interactions. More controversial, Lorenz asserted that aggression was an instinct shared by humans and animals. This was one of the causes of war, Lorenz argued. This idea, discussed in his 1963 book *On Aggression*, was met with angry resistance by many American social scientists.[2] The outcry against Lorenz's aggressiveness thesis was so severe that UNESCO officially endorsed a rebuttal against Lorenz entitled "The Seville Statement on Violence."[3]

Karl von Frisch studied the behavior of the honey bee. Like ants, this humble insect had nevertheless puzzled ethologists for years because of its complex social behavior, which no one thought was conscious. Frisch studied bees' ability to locate food sources and communicate their locations to other members of its hive. In the 1930s he showed that honey bees use a dance language to communicate food locations to other bees. Once a honey bee finds a feeding station, the bee recruits other bees to obtain the food from the same location. Some dances make use of tempo, and the angle of the hive, the sun, and the food station. Bees also "vote" for a new nesting site.

Other scientists also made critical contributions to the "new" behavioral sciences. In the 1950s, psychologist Harry Harlow found that infant monkeys, like human babies, required the comfort of parental contact. An "instinct" in a close relative to humans had apparently been found. This result was reported in Harlow's presidential address to the American Psychological Association in 1958, and was another step in the resuscitation of comparative psychology.[4] More evidence contrary to behaviorism accumulated when John Garcia of Berkeley finally succeeded in publishing his findings in the mid-1960s that rats seemed to have an instinctual avoidance of foods that may have recently made them sick. Garcia's work was initially rejected for publication since it seemed so at odds with the dominant belief that behavior avoidance was a result of conditioning.[5]

As the new thinking became increasingly legitimized, major theoretical advances were occurring. In the mid-1960s, scientists applied genetic theory to a persistent problem in biology: why did some species exhibit altruistic behavior?[6] If natural selection was essentially a selfish enterprise (described as the "the selfish gene" by Richard Dawkins) then why did some species, under certain circumstances, seem willing to sacrifice their energies, reproductive opportunities, or even their lives for others? William Hamilton considered this question in reference to bees. Why was it that bees and certain other insects (of the order *Hymenoptera*) had members in their colonies that devoted all of their energy and resources to care for the colony's queen, and were themselves never capable of reproduction? It seemed that those insects who had genes to code for such behaviors would never be able to reproduce, and so could not spread this gene throughout the species. Hamilton hit upon a likely answer, which he called "inclusive fitness theory." Since genes are inherited, not just one individual will carry it. Rather, a number of closely related individuals, all descended from the same ancestor, will carry copies of the same gene. The more closely two individuals are related, the more genes they should have in common. Such a relationship can be calculated mathematically: each parent will share half of their genes with each of their offspring; siblings will also share half of their genes; grandparents will share a quarter of their genes with each grandchild, and so on. Thus, since the worker bees shared half of their genes with the queen bee (she is the mother of all worker bees), they could increase the likelihood of the spread of these genes throughout the succeeding generations by aiding a close relative (i.e., their mother) in bearing offspring, rather than having their own.[7]

Hamilton's work inspired Robert Trivers, who in 1971 proposed that the reproductive behavior of each sex is likely to differ. In the advanced mammals, a female has relatively few offspring throughout her life, and must devote enormous care and attention to her offspring over the course of years. Thus, she would be most likely to have healthy offspring if she was very discriminating in choosing which males would have the opportunity to impregnate her. She would tend to choose males who were fertile, healthy, and could protect her and her young. A male, on the other hand, could have more healthy offspring if he copulated with as many healthy females as possible. In the rather cynical calculus of sexual selection, once a male fertilizes a female, she is constrained to bear the offspring regardless of the male's continued attentions or lack thereof. Therefore, a male would seek to mate with a female that he selected largely for her perceived fertility and healthiness. Males would tend to compete aggressively among themselves for the "privilege" of fertilizing a desirable female. Trivers called this sex-differentiated reproductive behavior "parental investment theory."[8]

As Donald Symons showed in 1979, the different mating strategies of each sex tends to utilize physical characteristics that humans think of as "beauty" and "sexiness" to indicate a potential mate's health. Both sexes seek these traits in their mates, though other factors might make one sex emphasize these desires more than the other. Mating behavior in both humans and other animals, reconsidered from the "new" perspective of sociobiology and evolutionary psychology, yielded further fascinating results.[9] Thereafter, as we will see in chapter 2, numerous studies were conducted to verify the relevance of these hypotheses for human behavior.

By the mid-1970s, the key points of sociobiology were again in place. The announcement of the "birth" of this new discipline was made by Edward O. Wilson in his masterpiece, *Sociobiology: The New Synthesis*. Henceforth, even in the midst of passionate controversy, sociobiology quickly grew in the elaboration of theory, the accumulation of evidence, and the acquisition of adherents.

Charles Darwin's concept of sexual selection was *the* most dramatic foundation for sociobiology, and perhaps even more so for evolutionary psychology. The biologist Geoffrey Miller has called the revival of sexual selection theory "swift, dramatic, and unique . . . . one of the fastest-growing and most exciting areas of evolutionary biology and animal behavior."[10]

He even seems to consider it the most important missing puzzle to the scientific understanding of human behavior:

> It is important to understand the peculiar history of sexual selection theory because virtually all of 20th century psychology, anthropology, paleontology, primatology, and cognitive science, as well as the social sciences and humanities, developed without recognizing that sexual selection could have played any important role in the evolution of the human body, the human mind, human behavior, or human culture. Since biologists have embraced sexual selection, we must face the possibility that most current theories of human behavior and culture are inadequate, because they may have vastly under-estimated the role of sexual competition, courtship, and mate choice in human affairs.[11]

Miller is not denying the simultaneous efficacy of natural selection. But perhaps the most important point of both is that organisms must reproduce relatively numerous, healthy offspring for evolution to occur. Therefore, for most animals, sex is the key to evolution.

After the publication of Wilson's book, advances in sociobiology and evolutionary psychology occurred with amazing rapidity. Work accomplished in the early 1980s demonstrated the mathematical feasibility of a corollary of sexual selection, "runaway selection," which was first conceived in 1930 by Ronald Aylmer Fisher.[12] Fisher had wondered if it was not possible for sexual selection to *overtake* natural selection, in a sense. Females of a particular species might fixate on the "sexual allure" of a particular male body part, and continually select males with increasingly extreme versions of this body part. In this scenario, the growth and maintenance of the body part requires significant biological resources, and might even lessen the animal's chances for survival. However, its power as a sexual stimulant more than compensates in evolutionary terms because it greatly increases the animal's reproductive potential. This process could continue until the survival of the species was threatened by the very difficulty in living with this "sexy" but cumbersome feature.[13] The formation of the male peacock's tail is usually considered as the classic example of runaway sexual selection. Behavioral experiments conducted in the 1980s verified that females of many species do prefer certain male traits that often "advertise" a male's fitness. In a sense, a male feature that requires large amounts of energy to produce or maintain advertises the male's health to the female. Amotz Zahavi has called this relationship the "handicap principle."[14]

Sociobiology also relied extensively on studies of nonhuman primate behavior as clues to the evolution of human behavior. As Edward Wilson explained, "By comparing man with other primate species, it might be possible to identify basic primate traits that lie beneath the surface and help to determine the configuration of man's higher social behavior."[15] Wilson believed that chimpanzees were similar enough to humans in terms of their genetic structure, their social life, and their mental properties as to be "nearly human" in some respects. If we assume that there is a correlation between similar genetic structure and similar behavior, this offers further evidence that some aspects of human social behavior rest upon a genetic foundation. There certainly would be no reason to expect similar social behavior in two closely related species if all behavior was dictated by culture.[16]

As a consequence of the rise of sociobiology, primatologists began to pay renewed attention to those aspects of nonhuman primate behavior that could bear upon the issues confronting sociobiologists. Studies concentrated on sexual behavior, the thinking processes (i.e., cognition) of chimpanzees, and the degree to which nonhuman primate societies demonstrated possession of the basic features of culture. Researchers found that males competed for females through a system of aggressive display and social hierarchy. Dominant males had sex more often, and with more females. Females, for their part, "cheated" on their male mates, and were essentially polyandrous (have more than one male sex partner) themselves. Chimpanzee communities do show some social behavioral characteristics different from human populations. For example, chimpanzees form more closely knit communities than do humans, and child rearing is even more of a female duty in chimpanzee societies than it has tended to be in most human societies. These behavioral differences suggested new clues to answering one of the most important questions in evolution: why human evolution took a path distinct from that of the other primates.

Understanding the extent of chimpanzee intelligence would help in pinpointing the differences in chimpanzee and human brain evolution. Following on the studies of Jane Goodall, William McGrew, and others, researchers found that chimpanzees were "smarter" than anyone had realized. For one thing, a number of experiments and observational studies revealed that it was very likely that chimpanzees were aware of their own existence as well as the existence of others. Unlike humans, however, they do not seem to be capable of wondering what other beings think about them as an individual.[17]

Some nonhuman primates also possess what could be described as "culture." They can make very crude tools with the intention of using them later. They can also teach others how to make these tools, or teach other skills. Japanese macaques, for example, have learned to use water to wash food, and have taught that skill to others in their community. The behavior continued in succeeding generations.[18]

These discoveries have obvious implications for explaining human behavior. In the past three decades, human beings have in a sense "rejoined" the animal world. Other primates are not as radically different from humans in behavior as was believed; and humans behave in some ways as "animals." This would be expected, given evolutionary theory.

After the sociobiological "breakthrough" of the mid-1970s, some psychologists sought to apply the new principles directly to the study of human behavior. Chapter 2 will examine their startling findings.

2

Recent Studies on Human Sexuality

Craig Stanford, while attending a seminar on primate societies, made the mistake of claiming that chimpanzee communities had cultures. The infuriated cultural anthropologists attending the seminar "fairly leaped across the seminar table" to verbally garrote Stanford. "Apes are mere animals," they lectured, "people alone possess culture. And only culture—not biology! Not evolution!—can explain humanity."[1]

The application of the concepts of sociobiology to human behavior, beginning most forcefully in the 1980s, gave rise to a new discipline: evolutionary psychology. It is not an exaggeration to say that evolutionary psychology has revolutionized the social sciences in the last twenty years, and has gained enormous public attention in the last ten. Feature articles on evolutionary psychology have appeared in *Time, Newsweek, Scientific American, Psychology Today*, and other popular publications.[2] The new discipline has also spawned a plethora of television documentaries. This chapter will outline the basic findings of the discipline from the early 1980s until 2007.

Evolutionary psychology focuses on the behaviors that have likely been inherited in the human species as a result of hundreds of thousands of years of evolution. From this viewpoint, the discipline attempts to imagine the environmental conditions and social relationships of humans in the distant past, long before the emergence of civilization. These conditions have guided the slow process of human evolution, and produced modern human beings. The earliest civilizations, existing about 6,000 years ago, are simply too recent to have substantially affected human heredity. In this view, we are the same beings as those who settled down in the first agricultural communities around 10,000 BCE. In some ways, the "animal" nature of humans may be much better adapted to the environment that existed prior to the emergence of agriculture and the beginnings of civilization. As

Desmond Morris put it, we must think in the context of the "Human Animal."[3]

If we did so, we would see a species in which there were considerable differences between the sexes. Males are usually about twenty percent larger and heavier than females. They have more upper-body strength, higher metabolic rates, more body hair, deeper voices, later sexual maturity, are more aggressive, tend to engage in riskier behaviors, and tend to die younger as a result.[4]

In reproduction, the females must contend with a lengthy period of gestation and years of nurturing their young. Also, human females are fertile only for about twenty years of their lifespan. With these demanding conditions, women cannot give birth to very many children compared to the females of many other species. If a woman is going to have descendants, each child she bears must have a fairly significant chance of survival to adulthood.

A man's reproductive success, on the other hand, is less closely tied to age and physical condition, since a man merely needs to generate and ejaculate sperm into a woman to impregnate her.[5] Since he could in theory impregnate hundreds of women, his chances of having descendants are less dependent on the survival of any one child.

These differences create an interesting sexual dynamic. Essentially, a woman will tend to leave more descendants if she can persuade the father of her children to provide her and her children resources to help them survive the conditions of a harsh world.[6] Men can provide food, shelter, defend territory, and protect their wives and children. They can tutor the children in social skills, hunting and performing work, and guide them in forming social contacts in their community, and in climbing the status hierarchy.

In 1991 Kristen Hawkes elaborated on this relationship with her "showoff hypothesis." A Paleolithic man might have hunted game not only to feed his family, but to feed others of his tribe, and thus display or even elevate his status in the community. This behavior would also account for the tendency among males to be more aggressive and engage in risk-taking to a greater degree than do females.[7]

Given the sexual dichotomy in humans, a woman is always certain that a child she bears is her own; a man is never as certain. If the woman has only one sex partner, rather than many, the man should feel more confident that her children are also his, and be more likely to provide those children all the resources at his disposal.[8]

David Buss has identified the various problems faced by prehistoric humans and the characteristics that would tend to be evolutionarily selected to cope with such problems. In general, we would expect that a woman would seek a mate who would invest his resources in her and her offspring. Men commanding such resources would probably have a high social status; be relatively mature in age; be large in size; have above-average physical strength and athletic ability; be intelligent; be ambitious and industrious; and command substantial material resources. These traits would be essential for having healthy children, and living long enough to help raise those children. Strong, brave men could also physically protect their families. Furthermore, women would wish mates who showed a willingness to invest their time and attention in caring for their wives and children. Such men would tend to be dependable and stable; kind; demonstrate love and commitment toward their families; and interact with their children in a positive manner. It would also be important simply to be able to tolerate the frequent presence of a husband, or even enjoy his company in an intimate, life-long family relationship. His ability to keep her amused might facilitate this companionate attraction. Thus a woman would most likely choose a spouse with a similar personality and values.[9]

Geoffrey Miller sought to explain the evolution of human intelligence through a female choosing to mate with a male who can demonstrate his intelligence to her or to her community. Miller calls this the "display hypothesis." Postulating a concept sure to hearten intellectuals and artists, Miller has suggested that men have created art, music, and dance essentially as elaborate courtship displays. The most intelligent men would likely produce the most creative and interesting art, and succeed in attracting the greatest number of mates. Miller has offered the provocative observation in support of this hypothesis that men in their prime reproductive years have produced the most art.[10]

A woman would want a man who is loyal because there is always the fear that he might choose to impregnate other women, and divert his resources to them and their children. She and her offspring could suffer and even die as a result.[11]

Taken together, it would make sense for a woman to choose carefully with whom to mate by accessing the health and resources of potential sexual partners. She would then mate with the healthiest, tallest, strongest, kindest, wealthiest, most interesting, and most loyal man available, and remain in a (at least apparent) monogamous relationship with that man.

Men have a different set of reproductive strategies. A man who could impregnate many healthy, reproductively fit women would be likely to have many healthy children. A youthful woman would tend to be healthier, and have more years in which she could bear children. This is necessary because a woman has a relatively short reproductive time span and must be young, healthy, and fit enough to carry a baby to term. A woman's reproductive health would be much more obvious through her appearance than would a man's reproductive value. A woman's physique is also very important. Her body should be built to bear children. Thus, a man should concentrate on seeking visual cues to a potential mate's age, health, and fertility. Together, we call these traits "beauty" and "sexiness." A beautiful, sexy woman would have a youthful appearance, smooth, clear skin; full, firm muscle tone; lustrous hair; optimum body fat; relatively wide hips; a flat stomach; firm breasts; a lack of physical deformity or signs of disease; regular and well-formed facial features; hairless lips; white teeth; clear eyes; a high forehead; a small chin and nose; an expressive face; a sensuous smell; a bouncy and youthful walk; and a high energy level.[12] Men who would prefer to have sex with a woman with wrinkles, gray hair, and stooped posture would have few or no children as opposed to a woman who "looked" young and healthy.[13] Heavy women or those with narrow figures; those who are unhealthy; or those who are "unattractive" would also generally not be as fertile.

Thus, men will tend to mate with as many attractive women as possible. If a society embraces monogamy, however, a man will have to devote more resources to securing his sole mate. In this case, he will be even more choosy in whom he selects.[14]

Darwin called these gender-specific sexual goals "intersexual selection." If indeed these sexual desires were generated through inheritable genes rather than through cultural influences, one would expect to find the same basic sexual desires throughout the vast majority of human societies.

Both sexes will tend to choose healthy, good-looking partners who are also enjoyable companions. Unhealthy mates might often be sick, and thus less able to care for their mate and children. In fact, the healthy mate might have to devote his or her time and attention to caring for the sick mate in addition to caring and providing for the children. If an unhealthy mate died, the condition of the surviving family members would be even more dire. Furthermore, a visibly unhealthy mate might carry a genetic disease or defects that could be passed on to his or her children.[15]

Kind, amicable, and communicative partners would tend to make a monogamous relationship more rewarding, and so encourage both partners to stay in the relationship. A stable, peaceful relationship would provide a more secure and protective environment for children, and enhance their likelihood of survival and acquisition of status and resources.

Another corollary to human sexual selection theory is that individuals will compete with others of their sex for the most desirable mates. Men would compete primarily in the acquisition and display of wealth; women would compete in the display of their physical attractiveness. This has been called "intrasexual selection."[16]

Of course, these hypotheses have little validity if they are not verified through experimental or observational evidence. The decades of the 1980s and 1990s saw numerous studies directed toward testing these hypotheses. In general, the results confirmed assumptions to a much greater extent than most scientists ever thought possible.

Researchers sought to determine if, in fact, men wished to mate with beautiful women, and women wished to mate with handsome, powerful, and caring men; if beauty and sexiness were correlated with fertility; if beautiful people attracted more members of the opposite sex for sexual relationships; and if individuals competed with members of their own sex for mates.

In short, almost all the results collected in the past thirty years verify the theoretical expectations of evolutionary psychology. Anthropologists have found that traditional hunter-gatherer societies, which we imagine are most like those that existed in prehistoric times, have clearly defined status hierarchies. Those men at the top of such hierarchies have access to more food, more territory, superior health care, more wives, and more children than do lower-status males. The abundant resources of a high-status male are shared with his family. The children, in their turn, inherit their father's status, at least in part, and the benefits that come with that status. The lower a male's rank in the status hierarchy, the less resources he commands for his family.[17]

Primatologists are interested in the behavioral parallels between non-human primates and humans, as would be expected by evolutionary psychology. Their observations also confirm evolutionary psychological theory. For example, primatologist Barbara Smuts has observed that baboon females form sexual "friendships" with certain males, who in turn protect them and their offspring from harm by other males.[18]

Social scientists have examined the extent to which the principles of evolutionary psychology seem to play out in more developed, modern societies. Since the mid-1980s, data has been accumulated through dozens of surveys, experiments in flirting or dating behavior, and studies of "in search of" ads in magazines. These have been conducted in dozens of countries with diverse cultures, to ensure that the conclusions drawn from these studies apply to humanity as a whole, making it more likely that any tendencies shown to exist are more likely a fundamental part of human nature, and thus presumably genetic in origin.

Researchers have concluded that, perhaps above all, human beings seek to mate with "beautiful" members of the opposite sex. As Nancy Etcoff puts it, "every culture is a 'beauty culture' . . . . I defy anyone to point to a society, any time in history or any place in the world, that wasn't preoccupied by beauty."[19] Men desire women who are beautiful; women desire to appear beautiful to attract men. To enhance and preserve their beauty, women spend more time and money caring for their appearance than do men.[20] Women also worry about their weight to a greater degree than do men.[21] Of course, women also desire "handsome" partners.

In 1994, Steve Gangestad and Randy Thornhill found that facial symmetry was one feature perceived as "beautiful." The more the left half of a person's face mirrored the right half, the more attractive that individual was perceived to be. If the faces of average-looking people of a particular sex are superimposed upon one other, asymmetrical faces or unusually shaped features tend to "average out"; the composite picture will look increasingly symmetrical and regular in features as more and more faces are superimposed. The result is an attractive composite photograph. Good looking, symmetrical people are highly sexually desired by the opposite sex. More symmetrical males had their first sexual experience three to four years earlier, on average, than did less symmetrical men.[22] In both sexes, the more symmetrical individuals had more sex partners. As we would by now expect, individuals with symmetrical faces and bodies also tend to be healthier.[23] This relationship holds because symmetrical individuals usually have normal fetal development and growth, good nutrition while growing up, and freedom from parasites.[24] Since all members of each sex compete for the best looking mates, men tend to marry women about as good looking as themselves, if all other attractive traits are factored out.[25]

Age has a notable affect on men's perceptions of a women's sexual desirability. Doug Jones has shown that men desire women who have a youthful, "heart-shaped" face with "childlike" facial features (a facial form known as neotany). These include large eyes, a small, upturned nose, and a small chin.[26]

For beauty and "sexiness" to enhance reproductive value, these traits must be correlated with healthiness, youthfulness, and fertility, especially in women. Randy Thornhill, Steve Gangestad, Karl Grammer, and others have found that this is indeed the case.[27] Women with such features usually have high levels of estrogen, and thus tend to be more fertile.[28] Also, neotanous features may trigger the nurturing sentiments men feel toward their young.[29] Youthful women have more years to reproduce than do older women. They are also healthier. Women are most fertile at age twenty-two, the very age at which men tend to find them the most attractive.[30]

Men perceive women with an "hourglass" figure as sexier, healthier, and more fertile than thin women, and much more so than heavy women.[31] In a study that garnered a great deal of public attention, Devendra Singh in 1993 determined that men from many cultures are most sexually attracted to women who have a waist-to-hip measurement ratio of about 0.7. These women also happen to be the most fertile, and have the fewest problems delivering children.[32] For every ten percent increase in the size of the waist relative to the hips, a woman's chances of conceiving in any given menstrual cycle declines by an amazing thirty percent.[33] Thus, a woman whose waist-to-hip ratio (WHR) is 0.8 is thirty percent less likely to get pregnant than a woman with a WHR of 0.7. Women with smaller waists relative to hips enter puberty sooner than do heavier women with larger waists. Also, heavier women with larger waists have more difficulty becoming pregnant, and have their first live birth at a later age than do more slender women with smaller waists.[34]

The rounded buttocks and softer curves of women,which men find "sexy," are due to the fact that adolescent females gain approximately thirty-five pounds of so-called reproductive fat around their hips and thighs as they mature. This weight is about equal to the 80,000 calories needed to sustain a pregnancy.[35]

As a woman ages, men tended to rapidly decrease their perception of her beauty.[36] This seems correlated with the fact that older women have higher levels of testosterone, making them less fertile. Higher testosterone levels in women can manifest itself as facial hair, which men find unattractive.[37] Women, on the other hand, do not judge older men as harshly. Some women even think that somewhat older men are more attractive.[38]

Some studies have shown that attractive females are more likely to marry than are less attractive females. In 1984, J. Richard Udry and Bruce K. Eckland found that women who were rated as unattractive in high school were significantly less likely to have married, fifteen years later, than were the most attractive women of their class. More attractive women also tended to marry at a younger age than less attractive women.[39] Particularly attractive women also tend to find husbands higher up on the social scale than would otherwise be expected.[40]

Women are just as choosy about the characteristics of their mates as are men. Studies show that women clearly value height, strength, social dominance, and athletic prowess in their mates to a much greater extent than do men.[41] One British study showed that tall men have more live-in partners, a lower chance of being childless, and a lower likelihood of having no significant mating relationship.[42] A number of studies have verified that tall, handsome men tend to acquire the most material resources, power, and status in society. For men, tallness and good looks operate directly in obtaining mates, as well as indirectly, by helping advance status, which itself tends to attract mates.[43]

Women also tend to be attracted to men with a large chin, strong jaw, and prominent cheekbones. Even though these features are physiologically difficult for men to produce, they are positively correlated with high levels of testosterone in such men, showing their high degree of fertility and disease resistance.[44]

Women are also attracted to men who they think will make good fathers. In one brilliant study by Peggy La Cerra, women rated a potential marriage partner as more attractive if the man was shown interacting playfully and lovingly with a child. A man ignoring a crying child was rated by women as less attractive. Interestingly enough, this relationship did not hold the other way around: men rating a woman for her attractiveness appeared indifferent to the extent to which she was caring for a child.[45] Other studies have determined that women prefer more socially dominant and masculine men to impregnate them, but men who are kinder, more generous, and more emotionally open to raise their children.[46] This dynamic might account for instances of extramarital relationships among women.[47]

Given men's anticipated reproductive strategies, it is no surprise that men tend to desire more sexual partners than do women. This sets up an interesting tension: over time, women have selected men who make good fathers and tend to be willing to engage in monogamy. However, these men also feel the contradictory impulse to be sexually promiscuous.[48]

Not surprisingly, men with attractive wives tend to be happier in their marriages; women who have wealthy and powerful husbands also are usually more satisfied with their marriages. As we would expect, both sexes place a very high value on having a healthy mate.[49]

In 1987, David Buss and Michael Barnes asked people in thirty-seven diverse countries to describe their ideal mate in terms of earning potential, industriousness, youth, and physical attractiveness. In their study, respondents rated the desirability of these traits in a spouse. The most important characteristics men desired in their wives, in descending order of importance, were: kindness and understanding; exciting personality; intelligence; physical attractiveness; good health; easygoing nature; creativity; desire for children; graduation from college; good earning capacity; good heredity; good housekeeping skills; and strong religious beliefs.[50]

Buss and Barnes found that women preferred men who were considerate, honest, dependable, kind, caring, exciting, intelligent, fond of children, well-liked by others, had a good earning capacity, were ambitious and career-oriented, had a good family background, and were healthy, tall and good-looking.[51] The most notable difference between male and female respondents is that men wanted wives who were particularly physically attractive; women found intelligence, good education, and good earning potential in their husbands particularly important.[52] The fact that these relationships hold true for most societies studied thus far supports the conclusion that some human behaviors are due to genetic inheritance rather than cultural values.[53]

Because of their polygamous desires, males tend to engage in more intense sexual competition than do females. In fact, in the mid-1990s it was discovered that even a man's sperm competes with the sperm of other men. Researchers found that there were basically two types of sperm: cone-shaped sperm, which rush to fertilize a woman's egg; and sperm with coiled tails, which tend to block the sperm from other men. Thus, if a woman had sex with two men within days of each other, the men's sperm would have to compete and essentially attempt to destroy the competitor's sperm for the opportunity to impregnate the woman.[54]

Men also experience sexual jealousy more than do women. This is because a man can never be absolutely certain that a woman's child is his. A man would tend to be less willing to invest his time, energy, and resources in another man's child; further, he would not wish his mate to do so at the expense of his true children. Therefore, human males are very protective of sexual access to their mates. Men tend to "guard" their mates from possible sexual contact with other men, perhaps even to the point of using physical violence against male competitors or assaulting his own mate if she is suspected of cuckolding him.[55]

Women, on the other hand, are always certain which children are theirs. Since a woman places a high value on a man's loyalty and his contribution of resources to her family, her feelings of jealousy are more provoked if her mate shows affection for another woman than if he is sexually unfaithful but does not have romantic feelings for his mistress.[56]

In one study supporting these different, gendered provocations of jealousy, sixty percent of the males felt greater distress imagining their mate's sexual infidelity, whereas eighty-three percent of the female respondents reported that they would be more distraught over their mate falling in love with another woman.[57]

The fact that women know which children are theirs, whereas a man must always be less certain, has been offered as a partial explanation as to why women tend to devote more of their time and attention to their children than will a man. Indeed, at least nine percent of children were actually fathered by a man other than their presumed biological father, according to estimates by R. Robin Baker and Mark A. Bellis.[58]

Both men and women compete with members of their own sex to attract mates. Sally Walters and Charles Bates Crawford have reported that women tend to compete with each other for mates by looking attractive; men compete through sports and through displaying resources.[59] Discussing other studies, Anne Campbell reports that women apply make-up to their most alluring facial features, dress sexy, tan their skin, paint their nails, and show concern for a potential partner in their quest to compete with each other for desirable mates.[60]

Issues of sexual aggression have attracted the attention of evolutionary psychologists, and provoked the most zealous criticism from opponents. If men are more aggressive and physically stronger than women, and desire to mate with more partners, men might use their greater physical strength in a sort of exchange with women: women might allow men to copulate with them in exchange for protection. Such a relationship has been observed in nonhuman primates.[61]

If reciprocal benefits are not possible for some reason, men might resort to violence to obtain sex from women.[62] In their controversial book *A Natural History of Rape*, Randy Thornhill and Craig Palmer argued that rape is not the product of inchoate male rage due to cultural reasons, but is the deplorable application of aggressive sexual violence toward attaining the goal of male reproduction: insemination of attractive young women.[63] As John Alcock has described this phenomenon:

> Rape is a cultural universal . . . [rape developed as males evolved an] ease of sexual arousal, the capacity for impersonal sex, the desire for sexual variety for variety's sake, a desire to control the sexuality of potential partners, and a willingness to employ coercive tactics to achieve copulations under some conditions. [Such behaviors] almost certainly contributed to an increase in the number of females inseminated by some ancestral males with a consequent increase in the number of offspring produced.[64]

Zoologists have also noted that rape occurs in many other animal species, especially among nonhuman primates.[65]

Furthermore, in 1995 Stephen T. Emlen suggested that a cause of family aggression might be the result of the different survival and reproductive vantage points of different members of a family. For example, if two single parents combined their families, each might neglect their stepchildren for the sake of their own children.[66]

Conclusion

As this work on sexual selection advanced, evolutionary psychologists also developed a branch of inquiry that attempts to explain the existence of altruism among organisms. Altruism would seem to be counter-selective: how could an organism benefit from giving up resources to another, while receiving nothing in return? It would seem that if an organism had genes that coded for such a behavior, it would be at a competitive disadvantage to the "selfish" recipient of the altruistic act, and hence fail to reproduce as readily as the recipient, or as would "self-centered" organisms that kept all of their resources to themselves.

Robert Trivers attempted to solve this puzzle in 1971. He suggested the theory of "reciprocal altruism": an organism might engage in altruistic acts if it would eventually receive similar "gifts" from the beneficiary when the original "donor" most needed it, at some later date. In this way, both the original donor and the eventual reciprocator would benefit more than if they had both remained "selfish." In this "gain in trade" scenario, a hunter with a kill too large for his family to consume might share the meat with a friend. Later, when the hunter had to feed his hungry family but lacked a kill, his friend, if more fortunate in the hunt, could reciprocate the original favor by sharing his own meat.[67]

This altruistic relationship can operate in situations other than procurement of food. Primatologists Kenneth Ronald Lambert Hall and Irven DeVore noted in 1965 that male baboon friends will help each other in fights with other males over sexual access to females. This aid would later be reciprocated. Velvet monkeys who take the time and effort to groom their friends will in turn be groomed by them later.[68]

Evolutionary psychologists have recognized that reciprocal altruism can only work if the population that utilizes it can recognize "cheaters" and punish or avoid them. Otherwise, an animal might "naturally" do favors for a friend who never returns the benefit. This would throw the "balance of trade" off, penalize the altruist in favor of the selfish individual, and destroy any evolutionary benefit for altruistic behavior. It would tend to disappear from the population. To avoid this, an animal would have to be able to recognize and remember other individuals; remember acts of altruism and reciprocation; communicate their needs to their friends; and be able to determine if the reciprocated act is equal in "value" to the original altruistic favor. Given the complexity of these acts, it is likely that only the higher animals are capable of reciprocal altruism. In fact it may be that the evolution of reciprocal altruism has contributed to brain development.[69]

*   *   *

The extensive theoretical and experimental progress of evolutionary psychology and sociobiology in the past thirty years is quite amazing. It is even more amazing that much of this "new" knowledge had been discovered long before.

Part II

The Birth of Sociobiology and Evolutionary Psychology

3

The Animal Nature of Humans

Thanks to recent researches in the United States, it was now certain that the races of man act in exactly the same way as the races of animals.

Giuffrida Ruggeri, "Our Work in Eugenics," p. 4

The question arises, therefore, whether sociology should take account of animal groups as well as of human groups. If we assume the evolution of the human from the subhuman there can be only one answer to this question:sociology must take animal societies into account.

Charles A. Ellwood, Sociology in Its Psychological Aspects, p. 18

Occasionally the question is raised whether it is permissible to speak of such a thing as animal sociology. . . . It appears to me that this view is based—to put it crudely—upon an overestimate of mankind and an underestimate of animals.

Friedrich Alverdes, The Psychology of Animals, p. 115

The change from monkey to man might well seem
a change for the worse to a monkey.

J.B.S. Haldane, The Causes of Evolution, p. 153

Human Instinctual Behavior

Charles Darwin is undoubtedly the most influential thinker in the history of biology. The constellation of ideas that composed his theory of evolution seem to lead inevitably to new discoveries in biology whenever they have been applied. What is perhaps even more astounding is that some of his most logical ideas were to a great extent eventually dismissed and forgotten. In the mid-nineteenth century, Darwin quite reasonably assumed that, since humans were animals, behavior patterns observed in the higher mammals must have some presence in human behavior as well. This includes instinctual behaviors. Essentially, instincts were thought to be inheritable behavior patterns that manifested themselves under certain conditions, and generally increased the animal's adaptation to its environment. In his 1872 book *Expression of Emotions in Man and Animals*, Darwin explained that those behaviors common to both the higher mammals and to man very likely had an instinctual component. Instincts originated in animals lower down the evolutionary ladder, in response to common environmental challenges. These instincts were inherited by succeeding species, including human beings.[1]

Darwin was revered by evolutionary psychologists and eugenicists in much the same way as an Old Testament prophet. Thus, the directions he advocated for future biological work were accorded unique status. Darwin's sanction of the belief in instinctual human behavior inspired a fierce devotion to the concept in the early evolutionary psychologists. They endeavored to determine which human behaviors had an instinctual basis.

In 1890, the great American philosopher and psychologist William James endorsed the belief that humans possessed instinctual behavior. James believed that instincts were set behavior patterns triggered by exposing the organism to specific environmental stimuli. Like other behaviors, instincts evolved through natural selection, and were nature's "solutions" to particular survival and reproductive problems. James asserted that humans had even more instincts than did other animals. These included instincts that bound people together, such as love and parenting; and those that provided protection against threats, such as pugnacity and fear of common dangers (e.g., heights, spiders, and so on).[2]

James's student Granville Stanley Hall discussed human instinct at greater length. Hall is most noted for becoming the first president of Clark College, in 1889, and aggressively promoting the school as a peerless scientific institution.[3] Hall was also a noted Christian eugenicist. He believed that "Jehovah's laws are at bottom those of eugenics."[4] As a psychologist, Hall established himself as America's first and foremost expert in adolescent psychology. He was keenly aware that biology had to go beyond its centuries-long fascination with morphology, and delve into a study of the biological bases of behavior. He was quite convinced that "the manifestations of instinct are just as differentiated and as persistent as those of morphology itself." Instincts such as "the will to live, love of offspring, fear, anger, jealousy, individual attachments, memory, attention, knowledge of locality, home-making instincts and senses" were shared by animals and humans alike, and their biological bases had to be investigated. Fortunately for this endeavor, Hall predicted, "studies of life and mind will henceforth be more and more inseparable just in proportion as genetic or evolutionary conceptions pervade our field."[5]

Edward Westermarck also made lasting contributions to human instinct theory. Westermarck was a Finnish anthropologist and eugenicist working at the University of London on the history and sociology of marriage.[6] He came to believe that "social institutions are to a very large extent primarily based on instincts." As with other aspects of biological evolution, the development of social instincts has been guided by the laws of natural selection.[7] In his most important example of a social instinct, Westermarck noted while studying Moroccan culture that children seem to have an innate aversion to sexual relations with individuals with whom they are raised. In most cases, these would be the children's parents and siblings. Westermarck also rejected the idea that humans had lived through a stage of promiscuous sexual relationships. They were instinctively motivated to form life-long mating pairs. These theses were presented in his masterpiece, *The History of Human Marriage*, in 1891.

The British psychologist William McDougall was virtually the anointed spokesperson for the concept of human instincts in the early twentieth century. McDougall's book *Social Psychology*, published in 1908, was the first work to attempt to demonstrate how hereditary instinct permeated human social life. Here McDougall confidently asserted: "the springs of all complex activities that make up the life of societies must be sought in the instincts"; basic biological instincts manifested themselves through increasingly complex social behaviors, as civilization developed.[8] For example, McDougall's instinct of "pugnacity" originally led to the "bodily combat of individuals." Pugnacity also gave rise to a variety of self-assertive impulses, such as revenge, rivalry, and moral indignation. These, in turn, led to the development of morality and law. In more recent times, pugnacity has acted through "more refined forms," such as litigation or national warfare.[9]

Like many of his generation, McDougall was generally content to describe instincts, without worrying overly as to how they were physiologically expressed. His rare attempts to elucidate the instinctual mechanism is rather reminiscent of the later work of Konrad Lorenz:

> In the typical case some sense-impression, or combination of sense-impressions, excites some perfectly definite behavior, some movement or train of movements which is the same in all individuals of the species and on all similar occasions; and in general the behavior so occasioned is of a kind either to promote the welfare of the individual animal or of the community to which he belongs, or to secure the perpetuation of the species.

For both animals and humans, instinctive behaviors accomplish this feat by aiding the organism in acquiring food, shelter, companionship, sex, the defeat of opponents, and ascendancy in social hierarchies.[10]

Throughout his life, McDougall maintained that human culture rested upon a hereditary, instinctive basis.[11] In a bid to spread his ideas abroad, McDougall left Oxford for a Harvard professorship in 1920, at a critical moment in the history of instinct theory in the United States. Until that time, the writings of leading American sociologists of the early twentieth century, such as Edward Ross, Charles Ellwood, Knight Dunlap, and Edward Thorndike echoed McDougall's description of instinctual behavior.[12] By 1920, however, the reaction against the notion that humans had instincts began to gain momentum. McDougall and his ideas became principal targets for attack.[13]

Before McDougall arrived in the United States, several British and American ornithologists were accomplishing important work on the biological basis of bird behavior, and seeing a connection to human behavior as well. In 1901 Edmund Selous made the first detailed observational record of a mutual courtship display in animals; in this case, the Great Crested Grebe. However, Julian Huxley's observations on the Grebe, made thirteen years later, gained much more attention. Huxley's elegant and engaging style, along with his insights into the evolution of courtship rituals, would move Konrad Lorenz a half century later to proclaim Huxley as one of the "Founding Fathers" of ethology (the study of animal behavior).[14] In his book on the Grebe, Huxley insisted that behaviors were as valid expressions of genetic traits guided by evolution as were morphological structures. In at least some cases, this proved true for human behaviors as well. While focusing on the Grebe in his book, Huxley occasionally pointed out that human beings also engaged in evolutionarily determined courtship rituals. Most human courtship actions are predetermined by heredity, Huxley declared. Though more complex and less fixed, human courtship rituals were "not very different" from those found in the Grebe.[15]

In another cross-species parallel, Huxley found flirtation rituals in both the Grebe and in humans curiously similar. In the Grebe, if one member of a mating pair was sexually excited while the other was lethargic or otherwise unavailable, the first bird might locate another of the opposite sex, and begin a pre-coitus courtship ritual with the tempting stranger. However, if the mate now sees its partner interacting with the stranger, the mate is inevitably "roused to action" and drives the stranger away. Thereupon the mates usually engage in a particularly strong courtship ritual. "Thus all the anger and jealousy is directed against the usurper, not against the mate—which again is distinctly human!"[16]

Huxley also believed that birds and humans shared the predilection to develop a sexual fascination with a new mate, which he equated with "falling in love." Mating was guided by "impulse, unanalysable fancies, individual predilection," though naturally at a much cruder level in birds than in humans. This also meant that both species exercised mate choice.[17]

At much the same time, Wallace Craig was accomplishing very important work on the instincts of appetites and aversions in doves. Craig's discovery of "appetitive behaviour" has been called "one of the most important theoretical contributions in the advance of Ethology."[18] Craig advanced the idea that desires and aversions actually follow a cyclical pattern in which a desire is built up, followed by achievement of the goal, rapid desensitization, and a return to a graduated reacquisition of the desire. He saw the same process operating in humans as well as in doves. "If a dove's cycles are determined largely by instinct, habit, physiological conditions, and not intelligence," Craig wrote, "so are some human cycles, as those of sleeping, eating, drinking, and sex." Even variations in the typical dove cycle can find an analogy in humans, albeit at a much more sophisticated level. If the nest of a dove is disturbed by a human, the dove may abandon it to build another. Similarly, in humans, "a man begins to build a house; when he has progressed far with the building he meets some horrible experience in it which 'turns him against' it, and nothing will induce him to proceed with the house; he abandons it and begins to build elsewhere." Craig also saw homesickness as a cyclical desire in humans. After reviewing the impact of human instincts on a series of activities and feelings as diverse as parenting, homesickness, and artistic appreciation, Craig concluded that "we, like the birds, are but little able to alter the course of our behavior cycles."[19] In another work written immediately after World War I, Craig compared the causes of human aggression to those found in animals. He decided that humans did not need to go to war; in animals, simple aggressive display was often enough to ward off combat.[20]

The English sociologist William Trotter was preoccupied with explaining the causes of the group behavior he witnessed during World War I. The "herding" instinct, prevalent in the higher mammals, was especially important in influencing human development, Trotter observed. The strength of the group increased the chance for each member's survival. Furthermore, those individuals with "distinctive" (and presumably beneficial) hereditary traits might be able to thrive only in a protective group setting. They would thus have the opportunity to pass on these traits to their descendants. Trotter believed that the evolution of the human capacity for speech and art could be explained through this process.[21]

Unfortunately, evolutionary psychologists had ample opportunity to ponder human aggressive instincts in the years to come. In the last days of the interwar era, in mid-August 1939, the geneticist Samuel J. Holmes might well have been thinking about the current geopolitical crises in terms of the destructive potential of aggressive animals possessing devastatingly powerful weapons. He saw this as a consequence of the inherent tension between civilization and human instincts. Holmes explained that human instinctual behavior had evolved to adapt to a prehistoric environment, in which "small quarrelsome clans . . . afforded ample opportunity for the exercise of the complementary traits of mutual aid and group pugnacity." However, civilized life emerged out of such a world with what constituted lightning rapidity on the evolutionary timescale. Human instincts could not possibly evolve fast enough to keep pace with the radically new challenges posed by civilization. Thus, Holmes warned, scientists had to realize that humans were biologically maladapted to modern life.[22] Fifty years later, evolutionary psychologists such as Robert Wright rediscovered the same paradox.[23]

In Germany, the gestalt-influenced zoologist Frederich Alverdes was particularly active in the on-going debate over instinctual behavior during the interwar period. Alverdes believed that instinctual actions were composed of two components: fixed behavioral patterns, and a "variable element" that was increasingly apparent as one progressed up the evolutionary ladder. In man, this "variable element" allowed for the production of the abstract elements of human culture.[24] This was not meant to imply that humans were in some sense removed from their instinctual drives; quite the contrary. As Alverdes explained, "The egoistic instincts, in particular, have a way of appearing less undisguised among men; but for all that human motives are often fundamentally quite as gross as those of any other mammal, only a whole arsenal of abstract ideas is employed to deck them out in more pleasing garb."[25] Like the other sociobiologists of his day, Alverdes saw the basic human institutions as the consequences of instinctual drives. Marriage, family, childcare, and social cohesion were all the results of "irrational" instincts, such as altruism.[26] One of Alverdes's more unusual observations was that an instinct of "curiosity" and "playfulness" existed in the primates. These were undeveloped in the anthropoid apes, and led to unproductive play; in humans, however, these energies were directed into useful activities, and gave rise to civilization itself.[27]

Though the work of Alverdes, Holmes, McDougall, and others may have achieved intuitive insights of some brilliance, the problem with these early discussions of instinct is their tendency to substitute assertion for evidence. Indeed, after Darwin, the first evolutionary psychologists acted as if the existence of human instinct was "self evident." In general, there was little attempt to clearly define what exactly constituted an "instinct," to demonstrate a link between alleged animal and human instincts, or to show, by means of hard experimental data, that the purported causes and effects of instincts actually existed.

The search for experimental data regarding instinctual behavior was left mainly to zoologists such as Jacques Loeb and Ivan Pavlov, who made remarkable discoveries tending to suggest that "instincts" were actually conditioned (learned) behavior.[28] Their early experimental results were sufficiently impressive to mislead later researchers into claiming that *all* human behaviors were in fact learned. As we will see, valiant efforts to correct the experimental deficiencies of the evolutionary psychologist school were made by Robert Yerkes, Samuel Holmes, Corrado Gini, and others in the 1920s and early 1930s. Their results should have reestablished evolutionary psychology on a much firmer foundation. However, the ascendancy of environmental behaviorism was, by then, nearly complete. Experiments in evolutionary psychology no longer seemed relevant to most social scientists.

Comparative Psychological and Social Behavior

The discussion of human instinct was intimately linked to theories of cognition in the animal world. Early sociobiologists and evolutionary psychologists attempted to chart the rising complexity of mental ability as one moved up the animal kingdom. Once again, Darwin led the way by presenting evidence that the higher mammals showed signs of possessing, in a rudimentary form, most of the higher mental abilities found fully developed in humans.[29] In the last years of the nineteenth century, George John Romanes sought to correlate the mental capabilities of animals up the scale of evolution with the mental development of the human child.[30]

However, the rather haphazard reliance on anecdotal evidence by Darwin and Romanes in their attempts to show the mental capabilities of animals wore thin for psychologists by the end of the nineteenth century. Philosophical speculations backed up by naturalistic observation seemed simply too naïve and "unscientific."[31] By the end of the nineteenth century, cutting-edge psychologists were focused on employing the techniques of experimentation and statistical analysis on their subjects. These new methodologies brought precision and rigor to studies of behavior, and hence came closer to discovering the "truth." The "hard" sciences had accomplished great strides through the application of statistical analysis to controlled experiments. Why were such techniques not possible in the case of animal behavior?[32]

By the 1920s, evolutionary psychologists recognized the critical importance of experimental animal psychology for providing a sort of psychological "missing link" between the obviously instinctive behavior of the lower animals and the apparently more elaborate behavior of humans.[33] Unlike the mid-twentieth century work of Lorenz, Tinbergen, and von Frisch, who concerned themselves mainly with non-mammalian animals,[34] the most vital research of the early-twentieth-century evolutionary psychologists concentrated on nonhuman primates.[35] Primate psychology occupied the attention of several important German psychologists, such as Alverdes, before World War I. In the United States, however, primate psychology was almost unheard of until the dogged efforts of Robert Mearns Yerkes to secure adequate funding for meaningful laboratory work, which he accomplished in the mid-1920s. Yerkes virtually revolutionized the study of chimpanzee psychology, always hoping to validate the "truths" of evolutionary psychology through his research. He began by asserting that much of chimpanzee behavior had an instinctive basis. If he could successfully demonstrate the similarity between chimpanzee and human behavior, he would be able to show that the human behaviors in question also had an instinctual component. Yerkes never abandoned this basic premise. However, the longer Yerkes studied chimpanzees, the more impressed he was by their capacity for *learned* behavior. In addition, Yerkes realized by 1930 that it was politically expedient to downplay his allegiance to evolutionary psychology.

When investigating the behavior of the primates and other higher mammals, evolutionary psychologists tended to focus on social behavior, with some consideration also given to cognitive behaviors (learning and problem solving). In general, animal social behaviors were easier to observe, and undoubtedly seemed more obviously related to human social behavior. William Trotter offered a number of clever insights into the behavior of larger animal social groups, which he called "herds." As discussed above, almost all animal sociologists recognized that one of the primary reasons for the existence of animal social groups was to offer greater protection to the individual members of the society than could be achieved alone. This phenomenon was, of course, judged to be instinctive. Trotter saw another advantage to herd organization: it allowed the individual animal to react to dangers with a greater degree of efficiency than was otherwise possible, and so allowed more time to search for food. Of necessity, an animal must have the cognitive ability to recognize who is a member of his herd, and who is an intruder. Vocalization evolved as an effective means of conveying warnings of intruders or other impending dangers to the other members of the social group, Trotter wrote. In man, the "herd instinct" manifests itself as a fear of loneliness, an acceptance of social norms, and a willingness to be led. Ethics are also derived from the herd instinct, in so far as those who deviate from the norm in their behavior will be seen as a threat to the larger group.[36]

Frederich Alverdes also was interested in the relationship of the individual to the social group. In both chimpanzees and in humans, potential danger could be signified by nonverbal communication as well as by vocalizations. Alverdes mentions an observation by the German primatologist Wolfgang Köhler, who studied chimpanzees at his experimental station on Tenerife Island around the time of World War I:

> Köhler could make every chimpanzee . . . look in exactly the same direction by suddenly behaving as through he were extremely frightened, and riveting a spell-bound gaze on a particular spot. All the chimpanzees would then run together as though thunderstruck, and stare at the same spot, even though nothing whatever was to be seen there. It is common knowledge that anybody can make the same experiment at any time with his fellow men by staring fixedly at a portion of the pavement, or into the air.

As humans developed, warning yells evolved into more complex vocalizations capable of conveying a vast range of information. This, of course, was the genesis of speech. Speech might still be used to signal group membership rather than convey specific information. Sometimes, Alverdes noted, human conversation was practiced not so much to impart information as simply to demonstrate a readiness to socialize with others.

Dancing was another significant form of communication in both animals and humans. Alverdes saw dancing as a mechanism to enhance self-display during courtship rituals. In humans, dancing had also become more generalized, as an aesthetic activity meant for a wide audience.[37] Much the same is said by modern evolutionary psychologists.[38]

Edward Westermarck was quite naturally interested in the insights primate behavior could throw on the human institutions surrounding marriage and family. The family exists, Westermarck explained, "because it has a tendency to preserve the next generation and thereby the species. The male not only stays with the female and young, but also takes care of them." This is a result of instincts acquired through the process of natural selection. Among primates, family formation was encouraged by the small number of young, born one at a time, and the long period of infancy.[39]

Robert Mearns Yerkes was unquestionably the most important researcher on primate behavior in the first half of the twentieth century. He has even been described as what amounts to the "father" of American animal psychology.[40] Born in rural southeastern Pennsylvania, Yerkes soon showed an interest in animal behavior; it was the first professional love of his life. The second was eugenics, which he also discovered at a young age. On a visit to the Zoological Laboratory at Harvard, Yerkes made a point of visiting the leading American eugenicist Charles Davenport. Yerkes told Davenport that he wished to devote his life to studying animal behavior.[41]

Yerkes went on to study with Davenport (and the eugenicist William Castle) at Harvard. Yerkes remained friends with Davenport and Castle for the rest of their lives.[42] Yerkes came away from Harvard convinced that animals inherited both instincts and consciousness. What occurred in the mind must be studied experimentally and understood by any biologist who hoped to understand behavior. And, for Yerkes, there was no essential difference between animal and human behavior.

Animal behavior, human behavior, and eugenics would always be Yerkes's trinity—they were one and inseparable, the three facets of sentient life centered on evolution. Yet, Yerkes's own ideas evolved over the course of his life, under the pressure of a changing society and scientific community. His "hard" eugenics and "hard" hereditary behaviorism, apparent in his early work, "softened" over time. A brief consideration of his career highlights this process, which was characteristic of the few evolutionary psychologists whose careers survived the environmentalist ascendancy in the nature–nurture debate during the 1930s.

In a 1912 letter to another fervent eugenicist, Henry H. Goddard, Yerkes revealed his early, radical eugenicism. Discussing a paper he was writing, Yerkes told Goddard that he intended to "particularly point out the danger in the present Infant Mortality movement if we stop with simply saving the lives of the infants and do not go further and attempt to prevent the birth of such as are likely to be defective."[43] Yerkes publicized his views concerning the relationship of animal behavior, human behavior, and eugenics the next year at a symposium on the study of human behavior held during a eugenics conference at Cold Spring Harbor on Long Island (home to the Eugenics Record Office). Yerkes proposed "to bring some of the experiences of the student of behavior of animals to bear upon the problems which the eugenic investigator meets." He noted that "human behavior . . . presents essentially the same kinds of problems as does the behavior of any other mammal" and so should be studied with the same methodology. Yerkes recognized the obvious limitations that manipulation of human subjects posed for the investigator. Thus, he proposed that nonhuman primates be substituted for humans in psychological experiments. A full-fledged center for the study of nonhumans should be built to facilitate such research. "I fully believe that this apparently round-about way to knowledge of the laws of our own behavior is in reality the most direct and desirable way," Yerkes explained.[44]

That year, in fact, Yerkes began work with orangutans at a research station near Santa Barbara. He was astounded by the ability of one orangutan, Julius, to move boxes and climb up on them in order to reach a banana. Working with Julius convinced Yerkes that nonhuman primates were capable of creating ideas ("ideation"), foresight, and planning.

His research was reported in his 1916 book, *The Mental Life of Monkeys and Apes.*[45]

Already, however, Yerkes had to contend with the growing split among his colleagues over the nature of behavior. Since 1912, Yerkes's friend and colleague John B. Watson had become the leading investigator and spokesman for the new psychological theory of "behaviorism." Behaviorism downplayed or even dismissed discussion of the internal mental processes, which it accused of being unscientific and smacking of "metaphysical speculation." Since only stimulus and response was observable and measurable, this learning relationship should be the focus of psychological investigation. Indeed, so intent did the "behaviorists" become on studying learning reflexes that many eventually concluded that "conditioning" was *the* human mental process; in other words, behavior was learned. Traditional psychological concepts such as "mind," "instinct," "consciousness" were at first doubted and later ridiculed.

This Yerkes could not do. Though he fully sympathized with the "new" twentieth century generation of psychologists in their insistence that experiment rather than speculation and anecdote had to be the foundation of theoretical psychology, the dismissal of virtually all cognitive processes seemed to him to be radically reductionist. Yerkes thought Watson's behaviorist thesis more an ideologically driven "confession of faith" than a well-grounded scientific doctrine. Science had to come to grips with consciousness, instinct, and mind, though in an experimentally verifiable manner.[46]

For perhaps a decade, the majority of American psychologists tended to agree with Yerkes. Indeed, he became the administrative head of American psychology in 1917, as president of the American Psychology Association.[47] This was the eve of the American entry into World War I. In the spring of that year, once the United States declared war on the Central Powers, Yerkes, Lewis Terman, Henry Goddard, Carl Brigham, and other psychologists patriotically offered their services to the U.S. war effort. Within a short time, Colonel Yerkes was commissioned as chief of the Army's Division of Psychology. He and his colleagues were asked to develop a battery of intelligence tests that could sort out the new draftees by intelligence, with the intention of assigning them duties requiring the appropriate intellectual capacity. The tests were administered to 1.75 million recruits.

The opportunity the tests afforded to "scientifically" measure the recruits' intelligence, categorize them by race and ethnicity, and then attempt to correlate the findings appealed to the less agreeable side of Yerkes's personality. Yerkes's group embarked on this project expecting that the tests would confirm the racial hierarchy of intellect then commonly accepted: whites of northwestern European descent would rank as most intelligent, with a decrease in intelligence as one moved geographically away from this region. Dark-skinned African Americans were expected to rank as the least intelligent of all American racial or ethnic groups. Since Yerkes's tests actually measured educational level and acculturation to white, middle-class American norms rather than intelligence, the psychologists found that the results exactly conformed to their expectations.[48] Yerkes confessed to Brigham that his tendency was to "emphasize on every occasion the importance of individual mental differences and also of racial peculiarities."[49]

After the war, Yerkes was perfectly happy to use his Army Intelligence data to further the cause of immigration restriction. This, too, he saw as a natural corollary to eugenics. In light of his scientific status and support of eugenics, in May 1919 Yerkes was inducted into the newly formed Galton Society for the Study of the Origin and Evolution of Man, the most exclusive eugenics society in the United States.[50] Yerkes and Brigham gave a presentation on racial and ethnic intelligence based on their Army IQ results during at least one Galton Society meeting, in 1922. Carl Brigham announced these conclusions in his 1923 book, *Study of American Intelligence*.[51]

Yerkes continued to put his primate research on hold over the course of the early 1920s, as he remained in Washington, D.C., to help direct the nascent National Research Council (NRC). His contact with Davenport and with Madison Grant, notorious as America's most virulent racist, continued while Yerkes sat on the Council: all three were active members of the Council's Division of Anthropology and Psychology. Davenport and Yerkes were also members of the Council's Eugenics Committee, organized in 1920 within its Division of Biology and Agriculture.[52]

In 1922 Yerkes became chairman of a new NRC Committee on the Scientific Problems of Human Migrations. During one of the initial meetings, the fateful decision was made to work more closely with the newly established Social Sciences Research Council (SSRC), which was also interested in migration issues. The two groups were radically different, however. Whereas physical anthropologists and eugenicists dominated the NRC Committee, behaviorist-oriented social scientists dominated the SSRC. Through these two groups' interactions, Yerkes got to experience, first hand, the antipathy that many social scientists were increasingly feeling toward the eugenics and anti-immigration agenda. As a man who always avoided conflict, Yerkes didn't enjoy the experience. He resigned from the NRC in 1925, and resumed his scientific research as a professor of psychobiology at Yale University.[53]

Several years before, in 1921, Yerkes had seen another opportunity to realize his dream of building a primate research station. He was approached that year by the president of the Carnegie Institute of Washington (CIW) and fellow Galton Society member, John C. Merriam, with a proposal to participate in a very small, select committee of scientists to investigate the bases of human behavior.[54] The Advisory Committee on Research on Human Behavior was eventually composed of John Merriam (a paleontologist), Edwin Conklin (a cytologist), Edward Thorndike (America's foremost child psychologist), Clark Wissler (an anthropologist who studied American Indians), Charles Davenport (a geneticist), Carl Seashore (a psychologist), and Robert Yerkes (a psychobiologist). Of these, all but Seashore were members of the Galton Society; all joined the Advisory Council of the American Eugenics Society upon its formation in 1923.[55]

Displaying the typical grandiose spirit of post–World War I America, the human behavior committee decided that their work should seek to ameliorate the "individual and racial, national and international, educational and social, hygienic, medical and political" problems that were "pressing for solution." With Merriam hinting that big money might be obtained to fund the committee's projects, proposals from individual members of the group flowed in thick and fast. Many of the proposals also took the opportunity to disparage their environmentalist opponents.

Davenport was excited about the possibility of doing "race psychology." He criticized the work on infant behavior conducted by the founder of behaviorism, John Watson, for Watson had neglected to obtain critical "information as to genetic constitution" of the children. After all, no other organ was as susceptible to hereditary differences as the brain, Davenport explained. And not all brains were created equal: scientists should know if they were studying "the brain of a male or female, of a Scandinavian or a Sicilian, of a student or a moron . . . a criminal or a farmer . . . of a philosopher or a 'promoter.' "

Thorndike was dismayed at the extent to which Watson's behaviorism had infected psychology. As he told the group: "The general drift of psychology is strongly toward explaining all learning as a matter of the formation of associations or connections or bonds between situations and responses, stimulus and reactions. Altho [sic] perhaps not a dozen psychologists would accept this simple formula as adequate, it has made enormous progress toward acceptance during the past fifteen years."

At one meeting it was pointed out, probably by Yerkes, that "the materials and methods of comparative and genetic psychology are at present tending toward neglect in spite of the fact that the scientific study of behavior and of the psychobiology of infra-human organisms during the past decade received more attention in this country than elsewhere in the world and advanced so rapidly that animal psychology came to be looked upon as an American science." This crisis could only be solved through the construction of a breeding and observation station for anthropoids. Yerkes made sure to emphasize to the committee the importance of primate study for understanding human psychology:

> It is eminently desirable that all studies of infra-human organisms, structurally and functionally, to man, should be made to contribute to the solution of our own intensely practical, medical, social and psychological problems. . . . There is every reason to suppose that the solution of many of the most interesting and pressing problems of experimental medicine, of human genetics, physiology, psychology, sociology and economics may be solved, at least in large measure, most directly and economically through the use of monkeys and anthropoid apes.

As if this wasn't enough, Yerkes, in conjunction with Carl Emil Seashore, wanted to also look at fetal and childhood behavior; the behavior of twins; transgenerational behavior; and the relationship of the mind to the brain and nervous system. Caught up in the moment, Carl Wissler dreamed of studying "the biological equipment of man for the culture of his group."

However, such all-encompassing proposals, by attempting to lead everywhere, actually threatened to lead nowhere. After a year of seemingly endless meetings and discussion about proposals, Yerkes confided in Thorndike that he was "becoming restless about planning and talking about research which I never see!"

Finally, the committee was able to pare down its goals and concentrate, as its members put it, on "the monkey, the baby, and the idiot." This would allow Yerkes to study chimpanzees; the Carnegie Institution of Washington's Department of Embryology to study the human fetus; and the Institution's Eugenics Record Office, under Davenport, to study the mentally disadvantaged. Since the latter two projects were already underway and funded by the CIW, it appeared that the Committee had done little more than waste its time.

Underlying the undoubted disappointment in the results was the simple fact that Merriam proved unable to deliver on the money he had suggested he could obtain. The measly $3,000 grant Merriam secured from the CIW's Board of Trustees to fund the Committee was insufficient to research much of anything.

In the end, only Yerkes got anywhere with his efforts. He had been determinedly pursuing research possibilities with chimpanzees, and had come across an eccentric Cuban dowager, Madame Rosalia Abreu, whose life hobby was the collection of primates at her estate, Quinta Palatino, near Havana. She agreed to allow Yerkes and his associate Harold C. Bingham to study her primates if they paid for their own expenses. As this apparently seemed to the CIW's Board of Trustees to be the only

Committee proposal worth much, Yerkes received a grant of $5,000 on February 4, 1924, for his Cuban expedition. The money was well spent: Yerkes acquired enough research material to write one of his most important works, *Almost Human*.

Yerkes made one of his most forceful arguments in favor of the human-like behavior of chimpanzees in this work. Before its publication, Yerkes explained to his colleagues on the Carnegie Behavior Committee that his book would ". . . present evidence of the emotions of fear, anger, resentment, hatred and retribution—emotions which in the main have to do with self-preservation and which are regarded as primarily 'selfish.' Of special interest . . . will be the indications of lasting resentment, seeming repentance, ability to detect sham, deceit, timidity, or dislike in persons, irritability or impatience, willfulness and whimsicalness. There are merely examples of certain varieties of human emotion which manifest themselves in the man-like apes."[56]

Yerkes made similar points in the text of the book. He asserted that the study of monkeys and anthropoid apes would "exhibit possible values . . . of these creatures . . . as means of deepening and making more highly serviceable our insight into the happenings and principles of [human] mental life, social relations, and educational effort." In the conclusion of this work, he was even more explicit:

> Much that has been written of the possibilities of profitable educational research with the primates applies similarly and with equal force to the investigation of social problems. Naturalistic studies of the primates should give us adequate working knowledge of their social relations and organizations, and of the chief factors of their social environment. There might appear, also, significant facts concerning social evolution and development, eugenic and euthenic practices or opportunities. And this information might enable us to see in a quite different light our own particular and perhaps peculiar social problems. We may not be eager to admit it, but it is none the less true that human social psychology and sociology are only slightly developed. May we not, perhaps, give them a great impetus by thoroughly acquainting ourselves with the facts and principles of social life and its resulting organization in our nearest of kin, the anthropoid apes?[57]

Yerkes agreed with the earlier observations of others that chimpanzee social group behavior was instinctive, and related to both the flocks and herds of lower animals, and to the family and social behavior of man.[58]

Yerkes's hopes that the CIW would fund a full-fledged primate research station would be disappointed. However, by 1929, Yale University and the Rockefeller Foundation finally came up with sufficient funds to realize Yerkes's dream: a chimpanzee research facility in Orange Park, Florida.[59]

Yerkes continued to study and write about chimpanzees even before his anthropoid station was funded. From 1925 to 1929, Yerkes was particularly interested in studying the cognitive processes of chimpanzees. His insights fascinated his colleagues in the Galton Society and other eugenicists, who saw the apparently close relationship between chimpanzee and human thought processes as an important affirmation of the core theses of evolutionary psychology. Yerkes obviously appreciated the support given his research by this group. He attended a number of the Galton Society meetings during this period, where he excitedly reported on his experimental findings (discussions of which were dutifully reported in the *Eugenical News*). Yerkes gave Henry Pratt Fairchild, future president of the American Eugenics Society, an autographed copy of *Almost Human* soon after its publication.[60]

As Yerkes explained to the Galton Society that year, the strict experimental protocols that he championed were relatively unknown before World War I. In his own work, conditions were presented to chimpanzees that allowed them to envision the creation and use of simple tools to reach an objective, usually food. The most important experiments involved splicing sticks together to provide a pole long enough to grasp food, and stacking boxes one on top of the other, creating a platform that would then be climbed by the chimpanzee and allowed him to grab food suspended from a height. Yerkes was convinced that the solutions to these problems came to the chimpanzees on their own. Thus, chimpanzees were capable of thought.[61]

Later that year, the *Eugenical News* discussed Yerkes's and Blanche W. Learned's book *Chimpanzee Intelligence and Its Vocal Expression*. Perhaps with the intention of countering John Watson's theories on childhood cognition, the author of the review emphasized the similarity between the cognitive processes of chimpanzees and human children, as discovered by Yerkes and other primate researchers. The newsletter reported that chimpanzees use the same learning processes as do human children, they just take longer. The author of the article also noted that chimpanzees demonstrated an "extreme and multifarious activity, like that of a child of three or four years . . ." Only with such recent work, the article observed, had "psychologists and educators . . . begun to appreciate the light that such researches are bound to throw upon the methods of learning, and especially the education, of children."[62]

The same issue of *Eugenical News* summarized another discussion on chimpanzee behavior held that month by the Galton Society. At this meeting Yerkes observed, as he would again several times in his career, that chimpanzees interacted with him in a manner which demonstrated their awareness of Yerkes as another entity.[63] Yerkes claimed that it was "intolerable" for a male chimpanzee he studied "to bear the displeasure of any one for whom he cared . . ." In a lecture a decade later, Yerkes announced that his laboratory was routinely making use of this significant aspect of chimpanzee intelligence to encourage them to cooperate willingly in the setting up and carrying out of sometimes uncomfortable experiments, provided that the subject knew and had good relations with the experimenter.[64] Yerkes also believed that chimpanzees had a sense of humor, and an extremely primitive level of speech.[65]

By 1926, the Galton Society was sufficiently impressed with the importance of primate research in providing evidence for evolutionary psychology that they proposed sponsoring the publication of Frederick Tilney and Henry Alsop Riley's manuscript, *Evidence of Evolution in the Brain of Primates*.[66] The Society continued to evince great interest in chimpanzee behavior over the course of its existence. In 1929, for example, Yerkes's colleague Harold C. Bingham, also of the Institute of Psychology at Yale, presented to the Society a paper on "Some Chimpanzee Adaptations that Are Fundamental in Human Achievements." Supporting Yerkes, Bingham also believed that chimpanzees showed a capacity for "insight." The box stacking and other experiments revealed that the chimpanzees, like humans, exhibited "planning, development of technique, anticipated solution, corrective judgments, reflective pauses, etc."[67]

Throughout this period, Yerkes continued to have a great interest in chimpanzee social behavior. Perhaps his most interesting research along these lines concerned chimpanzee grooming behavior, published in 1933. Yerkes explained that chimpanzees "groom" each other frequently, which involves carefully examining the hair and skin of a companion and removing parasites or other foreign objects. This process, Yerkes pointed out, "is conducive to the comfort, safety, or even the preservation of the individual" as well as an inducement for sex. Only primates have the visual acuity and manual dexterity essential for such an act.

Most importantly, Yerkes came to two "tentative" conclusions regarding grooming: it was a genetically inherited instinct found in all primates, and it was manifested in man. He cited several facts to substantiate his hypothesis that grooming was also a human instinct. He found that a similar behavior, "de-lousing," was practiced in many "primitive" human societies. Furthermore, he noted that children tended to groom one another without being taught to do so.

This evidence for an instinctual behavior in man allowed Yerkes to attack the (now) prevalent environmental behaviorists as directly as his non-confrontational manner permitted: "Anthropologists, it appears, have generally assumed that human delousing in primitive peoples is a cultural phenomenon. I have no where found factual basis for this assumption;

presumably it is lacking. If perchance the behavior pattern in point is primarily hereditary in chimpanzee, it would seem improbable that it should be cultural . . . in man or in any other type of primate in which it appears." Rather, "to the open-minded and curious student of the genetics of behavior, it irresistibly suggests the concept of recapitulation and tends to appear as the vestige of a phase of racial history in which grooming may have taken conspicuous place as social service." In the conclusion of this study, Yerkes boldly hypothesized that "grooming in chimpanzee . . . represents a genetically important pattern of primate social response, from which evolved incomparably useful forms of social service; . . . it represents a step in the socialization of primate behavior and stands as forerunner of human hair and skin dressing, nursing, medical and surgical treatment."[68]

Although Yerkes's grooming study was commented upon excitedly by other evolutionary psychologists, and noted in many subsequent works on animal behavior, Yerkes's claim to have demonstrated a primate instinct identifiable in man was a rear-guard action.[69] His study met with stony silence by the majority of his fellow psychologists, who by the mid-1930s probably considered such a claim to be quaint if not positively archaic. In the last years before his retirement, Yerkes found it expedient to turn his attention toward more purely physiological studies of the chimpanzee sensory and nervous systems, as well as advocacy of more humane care of laboratory chimpanzees.

Primate Sexual Behavior

As the battle over the relationship between human and other primate behavior reached its climax, several prominent biologists, who had until then judiciously kept themselves out of the melee, now felt called upon to lend their authority to the support of evolutionary psychology. Gerrit Miller, Jr., a highly respected zoologist working for the Smithsonian Institution's National Museum and a member of the Galton Society, was one of the most influential among them.[70] Miller wrote several important articles in the late 1920s and early 1930s on human and primate sexual behavior. The general tenor of these works was unmistakable: human sexual behavior was unquestionably a variant of typical primate sexual behavior. "The main characteristics of human sexual behavior, contrary to a widely prevalent belief, are not peculiar to man . . . most of them are shared by man with several non-human primates that have been subjected to comparative psychological study. In this department of his behavior man has invented nothing new." Just as a full understanding of human anatomy and physiology require comparison with other animals, human sexual psychology must rest upon a comparative foundation. The sexual behavior of the other primates, such as the chimpanzees, provided the most obvious reference point.[71] Miller's contention directly challenged the claims of the famous anthropologist Bronislaw Malinowski. Malinowski, a functional anthropologist allied with Franz Boas, deemphasized the relationship between human and animal behavior. For example, Malinowski argued that whereas animal "gregariousness" might be due to hereditary instinct, human "sociability" was a function of learned culture. Thus, the two were unrelated.[72] Furthermore, Malinowski did not believe that human sexuality operated based on instincts; nor did women have the opportunity to choose a mate; nor did fathers care about their children.[73]

Miller based his objections to Malinowski and the other environmentalists on several points. Humans, Miller claimed, show a greater proclivity to family life than do the other primates because, unlike other primate newborns, human babies need the care of both parents.[74] Nonhuman primates are sexually promiscuous; at the same time, human taboos everywhere seek to restrict sexual promiscuity, indicating that there is an obvious need to enforce monogamy to maintain civilized social organizations. "Nothing more is needed than a cursory acquaintance with the things which openly take place in the society to which we belong . . . to enable anyone to see the parallelism," Miller remarked.[75]

Certain sexual aberrations are also common to all primates. After reviewing the wide range of abnormal sexual behaviors found in common between humans and chimpanzees,[76] Miller wrote "the more conspicuous among those human forms of sexual behavior that are usually regarded as 'abnormal' or 'contrary to nature' are nothing more than little if at all modified aspects of traits so widely prevalent among primates that they must be recognized as parts of the racial heritage of every member of the order."[77]

According to Miller, the numerous commonalities between human and nonhuman primate sexual behavior led to a number of definitive conclusions:

(a) that it is now necessary for writers on the beginnings of human family and social systems to recognize that man does not possess a type of sexual psychology which differs radically from that of all other primates, (b) that is seems reasonable to believe that a stage of simian horde life with its attendant sexual promiscuity lies somewhere in the ancestry of the human social systems which exist today, and (c) that human social systems together with their defects will be best understood when they are interpreted as simian horde life incompletely modified to meet the needs of human culture . . .

With Malinowski's claims in mind, Miller countered:

> the common possession by men and monkeys of a type of sexual behavior which perfectly harmonizes with the needs of promiscuous life throws a heavy burden of proof on those which insist that the forerunners of existing men lived under a group formation totally different from that which appears to be the prevalent one among non-human primates. Furthermore, when human sexual behavior is looked at as it is and not as it is conventionally supposed to be we have little difficulty in detecting beneath the surface of the cultural structure unmistakable traces of the framework of the promiscuous horde.[78]

Like those of his colleagues, however, Miller's well-reasoned comparisons went largely ignored. He returned to studying other mammals and thereafter avoided reference to human behavior.

Perhaps the final major statement in this era supporting the use of comparative psychology to identify human instincts was made by Samuel J. Holmes. Holmes was professor of zoology at the University of California and an ardent eugenicist, joining the advisory council of the American Eugenics Society on its birth in 1923 and ascending to its presidency in 1938.[79]

In a speech to the Western Division of the American Association for the Advancement of Science in 1939, Holmes made a claim that surely must have stunned his audience for its audacious defiance of current scientific orthodoxy:

> If we wish to gain a proper insight into the native and uncamouflaged impulses of human beings there is perhaps no more instructive procedure than to study the group behavior of chimpanzees. There you will find the mutual sympathy, the group pugnacity, the egoism and the altruism which are so curiously blended in man.[80]

Probably Holmes's remarks were dismissed as merely the ravings of a deluded old man on the verge of retirement.[81] By now "the lamps had gone out" in the disciplines of sociobiology and evolutionary psychology. They would not be lit again for over thirty years.

4

Earlier Studies on Human Sexuality

Her hair should be voluminous like the tail of the peacock, long, reaching to the knees, and terminating in graceful curls; her nose should be like the bill of the hawk, and lips bright and red, like coral on the young leaf of the iron-tree. Her neck should be large and round, her chest capacious, her breasts firm and conical, like the yellow cocoa-nut, and her waist small— almost small enough to be clasped by the hand. Her hips should be wide; her limbs tapering; the soles of her feet without any hollow, and the surface of her body in general soft, delicate, smooth, and rounded, without the asperities of projecting bones and sinews. [Description of the ideal beautiful woman by a Sinhalese man]

John Davey, Account of the Interior of Ceylon, *p. 110*

Each sex is in a sense making, choosing, or keenly critical of secondary sexual qualities in the other.

G. Stanley Hall, Adolescence

Darwin and Sexual Selection

It appears that a sign of true brilliance is the ability to ask critical questions that precipitate more new questions than answers. Charles Darwin's questions about nature were so penetrating that biologists are still attempting to understand their implications. This chapter will consider the consequences of one of Darwin's most profound questions: what role does sex play in evolution? As Darwin pondered his theory of natural selection, he realized that it seemed to contradict some of the apparent oddities of nature. How could natural selection explain the extravagance of the male peacock's tail? It made the male much more visible to predators, cost enormous energy resources, and seemed to have no value in aiding the animal's chances of survival.

Darwin, however, zeroed in on the most essential feature of naturalistic evolution: any inheritable trait that increased the chances of an animal mating and bearing healthy offspring tended to increase its presence in a population over the course of generations. Even if a trait *decreased* the chances of survival, if the trait more than made up for its dangerous side effects by *increasing* the chances of mating and producing numerous healthy offspring, it would tend to spread in the population. Darwin called this process "sexual selection." Sexual selection might work if animals compete to attract mates, while at the same time struggling against same-sex rivals to obtain the most desirable mates. Indeed, we could even think of evolution as a process revolving around sex: the most "evolutionarily successful" animals are those that devote their lives, directly or indirectly, to producing the largest possible number of reproductively fit offspring.

Darwin realized that individuals were most likely to mate and have healthy offspring if they attracted mates by displaying outward signs of health and strength; if they offered "gifts" or were decorated with "orna-ments" to attract potential mates; if they had greater ability to detect and reach mates; if they possessed physical attributes that increased fertiliza-tion rates; and if they had weapons or strategies for repelling sexual com-petitors. His examination of nature revealed that sexual selection did indeed seem to occur almost everywhere. Animals, including humans, appeared to instinctually seek out and mate with the healthiest, most attractive, and most reproductively fit members of the opposite sex to which they could obtain access. He also noticed another interesting ten-dency: in most species, males seemed to compete more aggressively among themselves for mates than did females. Nevertheless, females still ultimately chose with whom to mate. This idea—that females were sexually empowered by nature—seemed quite unnatural to many late-nineteenth-century biologists.[1] As Darwin imagined the process working in humans:

> Preference on the part of women, steadily acting in any one direction, would ultimately affect the character of the tribe; for the women would generally choose not merely the handsomest men, but those who were at the same time best able to defend and support them. Such well endowed pairs would commonly rear a larger number of offspring than the less favored. The same result would obviously follow in a still more marked manner if there was selection on both sides; that is, if the more attractive and powerful men were to prefer, and were preferred by, the more attractive women. And this double form of selection seems actually to have occurred, especially during the earlier periods of our long history.[2]

Females had increased their beauty over the course of evolution by mating with the most attractive males. As Darwin put it, "If any man can in a short time give elegant carriage and beauty to his bantams, according to his standard of beauty, I can see no reason to doubt that female birds, by selecting during thousands of generations the most melodious or beautiful males, according to their standard of beauty, might produce a marked effect."[3]

Critics objected to Darwin's notion of sexual selection on several counts: it is teleological, in other words, it accounts for structures in terms of their final advantage. What could have caused the origin of such structures? In their most embryonic form, critics judged it doubtful that it would be a physical feature significant enough to attract the opposite sex. Sexual selection presumes that the female has "a certain level of aesthetic taste and critical power . . . and this not only very high and very scrupulous as to details, but remaining permanent as a standard of fashion from generation to generation, —large assumptions all, and scarcely verifiable in human experience." The necessary acuteness to beauty "is exhibited by no human being without both special aesthetic acuteness and special training." They found Darwin's belief that lower animals could have this level of aesthetic acuteness "too glaringly anthropomorphic." Finally, they reasoned that "Battle . . . constantly decides the question of pairing, and in cases where, by hypothesis, the female should have most choice, she has simply to yield to the victor."[4] Also, sexual selection seemed somewhat "indecent" to Victorian sensibilities.

Nevertheless, a small coterie of biologists were inspired by sexual selection theory. They systematically presented evidence against each of these objections over the course of the next half-century.

Interest in sexual selection built up only slowly in the last decades of the nineteenth century. In Britain, Darwin's cousin Francis Galton, the founder of eugenics, acknowledged the importance of sexual selection. Indeed, eugenics advocates a sort of "forced," conscious sexual selection, with the goal of "improving" the human species. Galton's follower Karl Peason conducted quite a few statistical studies examining several aspects of marriage; however, these dealt with actuarial data from married couples, rather than the sexual attraction that presumably brought them together.

It was not until the 1890s that serious consideration of the implications of human sexual selection became apparent. Several forces converged to make this intensified interest in sexual selection possible. Galton's ideas on eugenics had a much greater resonance in the scientific community than did Darwin's musings on sexual selection. Eugenics embodied the concept that human beings could be tested and ranked for various mental and physical traits. These included intelligence, "moral character," propensity to contract certain diseases, and almost any other characteristic one could imagine. Once individuals were categorized in terms of these rankings, those deemed "superior" would be encouraged to find a spouse among their peers and vigorously procreate (called "positive eugenics"). Those judged defective, on the other hand, would be discouraged from producing children (known as "negative eugenics"). Suggestions for encouraging the reproduction of superior couples and for discouraging the reproduction of those deemed inferior varied from educating the public about eugenic ideals to outright incarceration and sterilization of so-called defectives.

Furthermore, beginning around 1910, eugenicists became increasingly concerned that the intellectual elite of their societies was failing to marry, or at least to plentifully reproduce. These fears only accelerated over the ensuing decades, as the low birth rates of college-educated professionals continued to decline. The demographic consequence of World War I was an important catalyst for this trend. Hundreds of thousands of young men died because of the war, and so failed to reproduce. Some eugenicists, such as Roswell Johnson, saw it as their mission in life to counteract this trend. This led to a renewed interest in the forces that attract people to one another, allowing for marriage and procreation, and so increased scientific interest in human sexual selection.

Eugenicists such as Charles Darwin's youngest son, Leonard Darwin, promoted studies of sexual selection to better understand human motivations for mating. He reminded his audience at the 1923 annual meeting of the British Eugenics Education Society that, if both males and females prefer beautiful and healthy spouses, the most fit would mate with the most fit, and have relatively more numerous beautiful and healthy children.[5] The less attractive and sickly individuals would be forced to marry among themselves, if they married at all. The quantity of their children would be reduced accordingly. Over the course of generations, this process would become ever more accelerated.[6]

Leonard Darwin was disturbed, however, to note that in modern, "civilized" societies the supposedly most intelligent couples tended to have the fewest children, and the least intelligent the most children. This is where eugenic policies could come into play, he claimed. Knowledge of human sexual selection could be employed by scientific technocrats to guide human evolution, by segregating or sterilizing the inferior couples, thereby allowing the superior couples' children to predominate in the population.[7]

Leonard Darwin's status as a scientific expert rested mainly on the fact that he was the son of Charles Darwin. He was essentially a scientific dilettante, in the manner of Madison Grant or Frederick Osborn of the United States. Most of the early-twentieth-century eugenicists, however, belonged to the new generation of experimental biologists. They were convinced that "real science" had to be based on experimentation and statistical analysis.[8] To convince the public to embrace eugenics, they had to prove the validity of its claims using these methods. Thus the experimental eugenicists generated a vast array of studies, most often by collecting family genealogies and attempting to find mathematical correlations between the inheritance of various physical and mental traits.

Their task was made much easier by the rediscovery of Mendel's laws on inheritance, around the turn of the twentieth century. This marked the advent of the science of genetics. As soon as a few basic laws of inheritance had been worked out, by about 1910, eugenicists sought to utilize them in their quest to understand human inheritance. Some understood that human genetics was undoubtedly much more complicated than could be unraveled in a few years of experimentation on corn cobs and fruit flies. Others, such as Charles Davenport, rushed in with nearly religious fervor and produced mountains of studies on human inheritance. Often, these studies were characterized by poor data collection, glaring inconsistencies, and obvious statistical oversimplification. More cautious eugenicists were caught between their desire to validate their science and show loyalty to their eugenicist colleagues, and their obligation to produce sound and convincing research. Proving that human behavior was guided by hereditary drives such as those associated with sexual selection seemed to be an obvious starting point.

*   *   *

From the 1890s to the 1930s, approximately a dozen social scientists labored to develop the theoretical and (eventually) empirical basis for human sexual selection. Most of these scientists were well-known eugenicists. As a whole, they certainly gained more attention through their writings on eugenics than their writings on sexual selection and evolutionary psychology. Once their eugenic writings were disparaged and then ignored because of their bogus claims and disturbing recommendations, the eugenicists' solid work on evolutionary psychology was unfortunately also caught in the same conflagration.

The body of theoretical speculation and empirical research on evolutionary psychology conducted in the first forty years of the twentieth century is quite amazing. Many of the basic theoretical claims of evolutionary psychology made up to about 1990 had already been proposed several generations before. A considerable body of research produced in the earlier period also anticipated similar studies, and led to similar conclusions, in the last decades of the twentieth century. The reasons for this unfortunate neglect of the earlier work will be made clear in later chapters.

In both eras, the core principle that evolutionary psychologists sought to demonstrate was the relationship of physical beauty to sexual choice. In short, a "beautiful" body and face indicated health and reproductive fitness. Those who sought sex with such partners had a greater chance of impregnating them, and eventually producing healthy offspring.

Perhaps John Ferguson Nisbet was the first scientist after Darwin to elucidate the role of sexual selection in human behavior. Nisbet, a Scottish psychologist of the late nineteenth century, saw beauty as an unconscious signal of a potential mate's fitness, "that which is best adapted to a given purpose":

> That man or woman is beautiful who is best fitted for the conditions of life in which he or she is cast, and the preferences implanted in us are merely a device of nature's for furthering the interests of the species by the elimination of the worst types and the reproduction of the best.[9]

The Anglo-Finnish sociologist Edward Westermarck defined beauty as "the outward manifestation of physical perfection or fitness." Beautiful "bodily qualities" act as sexual stimulants for both sexes, he continued. "They are expressions of vitality, vigour, and health, or are closely connected with propagation." This instinctive preference for beautiful sexual partners had developed over the eons through natural selection.[10]

Westermarck's interpretation was universally accepted among evolutionary psychologists. They described the evolutionary significance of beauty in similar language. Knight Dunlap explained the concept in quite unmistakable terms:

> ... the most beautiful woman, the handsomest man, are the persons we would choose to be coparents of our children, if we considered nothing but the highest mental and physical welfare of these children.... Human beauty ... is a sign of fitness for parenthood; fitness to propagate children who shall be, in high degree, able to hold their own in the mental and physical struggle with nature and with their human competitors. It is the sign which is intuitively recognized by the race and upon which the process of sexual selection is based.[11]

Because both sexes had to compete to obtain the best-looking mates possible, and not everyone succeeded in mating, only the most beautiful men and women most likely mated, and attempted to do so with the most beautiful members of the opposite sex. John Nisbet discussed this tendency as the desire for a good-looking man "to seek a partner as good or better looking than himself."[12] This phenomenon was known as "preferential mating." The process, over time, resulted in evolutionary change in favor of those characteristics considered beautiful. To put it simply: because of sexual selection, the human species became more beautiful over time.[13] This process occurred regardless if mate selection was conscious or unconscious.[14]

As discussed in Chapter 1, Ronald Aylmer Fisher in 1930 suggested an interesting consequence of the process of sexual selection based on attractiveness. Over time, continued sexual attraction for certain physical traits might produce animals with highly exaggerated features, such as moose horns. Obviously, this "runaway sexual selection" could proceed until the animal's chances of survival became so diminished that its ability to reproduce was also adversely affected. Though a brilliant hypothesis, the eminent biologist Julian Huxley, who by 1938 had joined the now victorious environmental behaviorist camp, dismissed Fisher's suggestion. "Runaway sexual selection" was not seriously discussed again for another forty years.[15]

The first evolutionary psychologists were concerned about much more fundamental matters. Chief among them was the question: what physical characteristics constitute beauty, and how exactly do they act to enhance health and fertility? Interestingly, Charles Darwin refused to believe that there were universal standards of beauty among different cultures. He concluded: "It is certainly not true that there is in the mind of many any universal standard of beauty with respect to the human body."[16] Later evolutionary psychologists, however, disagreed. They enumerated the physical characteristics of human beauty, and over time sought to prove that these characteristics were indeed considered beautiful by potential mates the world over.[17] Carl Easton Williams, fusing the concept of a universal standard of beauty with the American racism characteristic of his era, claimed that "if we could see the features of some of the dark skinned races faithfully reproduced with out own fair colorings we should find that many of them were really good looking from our own standards."[18]

Almost all of the human physical features found to constitute beauty, as demonstrated by evolutionary psychologists in the 1980s, were known to their predecessors. Francis Galton first noticed that facial features "average" in form were more beautiful than were abnormal variations. He discovered this by superimposing images of various individuals, one upon the other, and examining the composite effect. As Galton described the phenomenon, "all composites are better looking than their components, because the averaged portrait of many persons is free from the irregularities that variously blemish the looks of each of them."[19] Unfortunately, it does not seem that Galton's discovery was incorporated into the literature of the early evolutionary psychologists. Rather, science had to await the similar studies of Judith Langlois and Lori Roggman in 1990 to further explore the ramifications of Galton's findings.[20]

The early-twentieth-century evolutionary psychologists, however, were fascinated with describing other physical signs of beauty. While mentioning traits found to be beautiful by both sexes, they generally concentrated on describing beauty in the eyes of each sex, much as would the later evolutionary psychologists.

Early on, John Nisbet saw "regularity of features" as an essential characteristic of beauty. Such regularity implied health and adaptive fitness. Conversely, "physical disfigurements that are permanent, perhaps a hereditary defect, are particularly unattractive."[21]

Knight Dunlap added to Nisbet's observations. Dunlap was one of the more curious contributors to this research. Dunlap was a professor of experimental psychology at Johns Hopkins University, where he worked closely with the leading behaviorist of the time, John Watson. Like Watson, Dunlap eventually condemned the supposedly anachronistic concept of human instinct in favor of behaviorism.[22] However, he was a passionate eugenicist in his earlier years. Dunlap was one of the original members of the advisory council of the American Eugenics Society. In 1920 he voiced concern over the eugenic consequences of immigration.[23] He also made his most radical statements with regard to eugenic sterilization that year. Dunlap did not tolerate the reproduction of the unfit:

> Feeble-mindedness, hereditary insanity, and hereditary criminal tendencies (if such occur) should be nipped in all the buds they show. Individuals showing these traits definitely should not be allowed to reproduce. Diseases and organic weaknesses which are transmissible to offspring (if there be such diseases) should come under the same rigid ban.[24]

His chilling solution to the problem of "dysgenic individuals" can only remind us of the Nazi "euthanasia" that would be carried out nineteen years later:

> It seems not only useless but dangerous to preserve the incurably insane and the lower grades of feebleminded, even when we consider the case from the individualistic point of view. When we estimate what the personal labor put into asylums and into institutions for the feeble-minded, might accomplish if expended in the poorer districts of our cities in teaching the children who will be the parents of a large fraction of the next generation of citizens, how to work and play, it seems a pity that we cannot asphyxiate the hopelessly insane and feeble-minded as kindly as we do stray dogs and cats.[25]

However, besides preoccupying himself with such obscene ideas, Dunlap interested himself in the late 1910s in human sexual selection. He outlined his thoughts on the evolution of human beauty in a 1917 address to the

Southern Society of Philosophy and Psychology, and published them three years later in a book, *Personal Beauty and Racial Betterment.* Dunlap extensively described those features considered beautiful in both sexes. These included a "well-developed chin," "graceful movement," "tactually felt smoothness of skin and firmness of muscle and glossiness of hair." Form, coloration, and a pleasant voice were also important features.[26]

As scientists would once again in the 1990s, Dunlap carefully considered the importance of body fat for beauty and health:

> A certain amount of fatty tissue is normal, and is essential for the health of the individual. Fat constitutes a store of reserve material, which may be drawn on in time of unusual need; and without it endurance is limited. This reserve store is probably not so important at present as it was in primitive times, when man lived in a hand-to-mouth way, uncertain today what the food supply would be day after tomorrow. On the other hand, beyond a certain amount, fat is an encumbrance, impeding the operation of many organs, and thus limiting the efficiency of the individual, and also is in itself a symptom of faulty organic functioning of some kind. We are not surprised therefore to find that beauty demands just the right amount of leanness; just the degree which is found in the most vigorous individual.[27]

Women's extra body fat served as a reserve supply of calories, "laid up against the heavy demands which are made by child-bearing . . ."[28]

Dunlap was aware that signs of ill-health or abnormalities negatively affected perceptions of beauty. These included "*significant deviation* from the average" for a given physical feature; or "*signs of disease, deformity or weakness.*" Furthermore, physical characteristics most often identified with one sex would not be perceived as beautiful if they appeared in the opposite sex. Thus, "the masculine woman or the effeminate man would not appear beautiful to normal people."[29]

Paul Popenoe and Roswell Johnson added "good complexion, good teeth and medium weight" to the list of desirable features. Carl Easton Williams mentioned facial symmetry as another important component of beauty.[30] It was agreed that all of these features indicated health.[31]

The earliest studies to confirm suspected attributes of beauty for both sexes were crude and not particularly revealing. Havelock Ellis, a British eugenicist and eminent sexologist, reported in 1905 that he had performed a "small statistical study" to show that individuals were in general attracted to tall members of the opposite sex.[32]

G. Stanley Hall, in his research on adolescent sexuality in the first years of the twentieth century, submitted a comprehensive questionnaire of attractive traits to adolescents. They were instructed to choose those traits they thought were the most attractive in members of the opposite sex, and those that were least attractive. Hall found that traits most admired were, in order:

> attractive eyes, hair, stature and size, feet, brows, complexion, cheeks, form of head, throat, ears, chin, hands, neck, nose, nails and even fingers, and shape of face. Pronounced dislikes included prominent, deep set eyes, full necks, ears that stood out, brows that met, broad and long feet, high cheekbones, light eyes, large nose, small stature, long neck or teeth, bushy brows, pimples, and red hair.[33]

In a newspaper interview that year, Hall added good teeth and broad shoulders in males to the list of likes.[34] These early studies were noted again, some years later, by Paul Popenoe and Roswell Johnson.[35]

A year after Hall published his findings on adolescent attractiveness, the popular magazine *Health* ran a series of articles entitled "The Woman Beautiful" by the physician C. Gilbert Percival. After devoting "years of study" to the perfect human form as indicated by ancient and modern sculpture as well as other sources of data, Percival announced those measurements characteristic of the "perfect physique." Women, for example, would ideally be 5 feet 6 inches tall, weigh 124 pounds, and have a bust of 36 inches and a waist of 24 inches. The ideal man would be 5 feet 10 inches in height, weigh 150 pounds, and have a bust of 36 inches and a waist of 32 inches.[36] The perfect woman's form, Percival explained, was "lithe and sylph-like." However, even heavier women were still attractive, as long as they had "harmonious proportions."[37]

One interesting attribute of personal beauty intrigued several scholars over time: people tend to marry others of approximately equal attractiveness. Havelock Ellis suggested this correlation in 1905, when he noted a study conducted by Hermann Fol, who found that there was a remarkable similarity in the facial features of married couples.[38]

In 1931, the Italian demographer Carlalberto Grillenzoni also found that most marriages were between people of "equal beauty ratings."[39] This finding would be mentioned again in 1938, in the English-language journal *Human Biology*.[40]

Other personal characteristics besides physical beauty could enhance one's sexual attractiveness. For example, personal adornment, perfumes, "sounds," and "erotic dances" could heighten sexual passions. William McDougall, who wrote extensively on sexual instincts in the 1919 edition of his *Introduction to Social Psychology*, discussed the relationship of clothing to sexual expression:

> In many subtle ways woman's dress manages, without transgressing the limits set by convention, to draw attention to and to accentuate her secondary

sex characters; and that it serves at the same time to conceal the body is also obvious. And many masculine fashions of dress serve the same two opposed purposes.[41]

Milo Hastings explained the difference in degree of ornamentation of each sex as a function of that sexes' mating opportunity. The greater the difficulty for one of the sexes to find a mate, the more likely that sex sported extravagant ornamentation.[42]

McDougall, Ellis, Westermarck, and others noted that sexual cries and "erotic" dances also could be found in some mammals.[43] As discussed above, modern researchers have confirmed these findings. These early researchers in sexual selection, most of whom were also eugenicists, were pleased to find that intellect could also enhance physical attractiveness in both men and women.[44]

One of the more interesting, and unusual, sources for information on beauty and human sexual selection was a popular magazine of the era: *Physical Culture*. The fitness and health enthusiast Bernarr Macfadden founded the magazine in 1899 to publicize his passionate views on healthy living. He encouraged vigorous and frequent physical exercise, consumption of vegetarian health food, sexual activity, enjoyable marriage, birth control, and eugenics. Though Macfadden lived a very idiosyncratic life, he was no fool: *Physical Culture* reached a circulation of half a million by the time of World War I. He would build upon the magazine's success to create a vast publishing and healthy living business empire.[45]

Physical Culture was imbued with Macfadden's fascination with beauty and sexual attraction. The magazine took an entirely Darwinian view of human nature, and discussed human sexual selection to an extent virtually unparalleled in any other popular publication of the time. All of *Physical Culture's* writers conformed to one interpretation of human sexuality: beauty was the outward manifestation of health, vitality, and fertility. Those who selected their mates according to the criteria of beauty as well as superior personality traits were fulfilling their eugenic duty. MacFadden advised his male readers to select women who possessed a healthy sexual instinct, were strong and vigorous, and were "well nourished and well rounded." Furthermore, "It is always well to investigate the family characteristics that are going to be implanted in your prospective children. . . . It is a good plan to find out what is in the 'stock,' and then be governed accordingly." Marry a woman, Macfadden instructed, "because you are convinced that she would be the mother of strong, splendid children."[46] In support of these positions, the writings of Havelock Ellis were frequently quoted or paraphrased in *Physical Culture's* articles, and Ellis frequently contributed to the magazine. As will be shown below, *Physical Culture* made important contributions to the fund of data drawn upon by the early evolutionary psychologists.

Traits Men Sought in Women

The early evolutionary psychologists were fully aware that those physical traits men desired in women were also the very traits that indicated health and fecundity. Indeed, Edward Westermarck contended that men consciously sought out women who were likely to be fecund. In many societies, a woman's status partly depended on the number of children she bore.[47]

William McDougall recognized that men were attracted to women with "childlike" facial features. Similarly, an editor of *Physical Culture*, Carl Easton Williams, recognized that one form of female beauty was the "sweet round face and the baby stare."[48] In the 1920s the infamous German racial hygienist Fritz Lenz observed that some women had "infantile" facial features that persisted even as they approached thirty years of age. Lenz noted that this child-like facial form "is known to exercise much fascination over many men."[49]

At the time, these observers did not offer a theoretical explanation for this phenomenon. Still, the assertion that men preferred "baby-faced" women anticipated the similar conclusion by Irenäus Eibl-Eibesfeldt and Victor Johnson in the 1990s. Perhaps over the millennia, men inclined to nurture children sexually selected women with neotanous features, thus reinforcing these traits. Women who appeared "child-like" would themselves benefit from the nurturing response of their mates, as would their children.

The early researchers into human sexual selection found that virtually all other female physical characteristics that sexually stimulated men were critical for female reproductive health. As Devendra Singh rediscovered in 1990, primary among them was a narrow waist paired with well-rounded hips and buttocks, the so-called waist-to-hip ratio (WHR), a concept that became a household term in the 1990s.

John Nisbet believed that "the great function of women is to bear children." Thus, men sought out women who had physical features best adapted to this task. "A full development of hips and bosom meaning easy *accouchements* and plenty of nourishment for the offspring, is everywhere an accepted element of female beauty. At the same time stoutness is deplored, because it implies sterility."[50]

Havelock Ellis explained that the large hips and buttocks of attractive women represent "the most decided structural deviation of the feminine type from the masculine, a deviation demanded by the reproductive function of women, and in the admiration it arouses sexual selection is thus working in line with natural selection." He also noted that "the purposeful vibration or cultivation of carriage" by women was meant to display the hips and buttocks to attract potential sexual mates. Full, attractive breasts, a long neck, attractive hair and lips, good color, clear skin, normal weight, and "innumerable other qualities of minor saliency" are also admired by men.[51] All these observations led to only one conclusion:

> The beautiful woman is . . . obviously best fitted to bear children and to suckle them. These two physical characters, indeed, since they represent aptitude for the two essential acts of motherhood, must necessarily tend to be regarded as beautiful among all peoples and in all stages of culture . . .[52]

Marion Malcolm paraphrased Ellis's observations in a 1914 issue of *Physical Culture*.[53]

Appropriately for *Physical Culture*, Carl Easton Williams reminded his readers that "the workings of the principle of natural selection are invariable . . . the health and vigor of the individual woman has much to do with the impression of her beauty" the world over.[54]

The writers of *Physical Culture* created a number of measures of physical beauty. In fact, one of the magazine's most popular features was the publication of the photographs, physical measurements, and accompanying statistical analyses of reader's submissions for "most physically fit" competitions held monthly by the magazine.

Milo Hastings frequently contributed to *Physical Culture* on the subject of physical beauty. A noted scientist, nutritionist, and author, Hastings developed a measure of feminine beauty he called the "ratio of sex differentiation."[55] It sought to statistically emphasize those parts of the female body that were usually either thicker, or thinner, than the male counterpart. Hastings calculated the ratio as equal to the measure of the (bust + hips + thigh)/(neck + waist + wrist + ankle + 1/2 height). Hastings decided that the ideal ratio of sex differentiation for women was 113 percent. The average American woman (based on measurements of the students of Wellesley College) was only 106.1 percent. The Venus de Medici fared even worse, however: she attained only a paltry 105.4 percent on Hastings's ratio. Hastings might have been heartened to hear that, using the Wellesley College data he provided, the average American woman had a 0.7 WHR.[56]

Marion Malcolm found that American women's WHR was even smaller in years past. Using data collected over a forty-year period by "one of the oldest clothing and pattern manufacturers in the United States," she calculated the average waist and hip measurements of American women, by bust measurement. Averaging the WHR for busts of all sizes yields a ratio of 0.64.[57]

Many observers noted that women's clothing was obviously designed to accentuate their physical attributes in ways to attract men. In 1899, the French physician Charles Féré claimed: "Women in even the most advanced societies retain a marked tendency to emphasize their sexual characteristics through clothing: for example, the corset makes the breasts and hips appear more prominent . . . Their attachment to dress shows that women count more on physical means than on intellectual or moral means to attract men."[58] The English sociologist Walter M. Gallichan deplored the fact that "everywhere women shape themselves by artifice in the manner most approved by the males." Besides restrictive clothing that emphasized the smallness of the waist, this also included high-heeled shoes that accentuated the swaying of the hips.[59] Bernarr Macfadden would have sincerely approved of these sentiments. He is credited as one of the most important voices for convincing women to abandon the corset.[60]

Milo Hastings, in one of his articles for *Physical Culture*, discussed body ornamentation and sexual selection. His reliance on the explanatory powers of Darwinian sexual selection is unmistakable: "Those things that have to do with beauty or ornamentation in animal life are almost always explained by sexual selection, and when the modern thinker questions the whys of human decoration he inevitably looks for his causes in this all-pervading force of amorous attraction."[61] Self-inflicted scars, tattoos, elaborate hairstyles, and elegant clothing (or the lack thereof) were adornments meant to attract the opposite sex. As for jewelry: "on what lesser passion should one creature sacrifice the life and labor of another to hang a bauble on a third creature's breast?"[62]

Men were also interested in mating with women who displayed certain personality traits, along with physical beauty. Walter Gallichan accepted that beauty might be the most attractive trait in a woman when it came to finding a mate. However, men also sought out women notable for their "kindness, ardor, sympathy, intelligence, and capability." Vivacity, even temperedness, and tranquility were also desirable.[63] Carl Easton Williams assured the readers of *Physical Culture* that "Without doubt personality, as the expression of mental qualities, has as much to do in the matter of selection as physical charms."[64]

Realizing that acceptance of evolutionary psychology might well rest upon the scientific validity of these conclusions, these scientists set out to conduct a number of statistical studies to prove their claims.

Preferential Mating Studies

Preferential mating studies had to overcome one major obstacle. Measuring the physical attractiveness of women was quite easy: disinterested male judges could rate photographs of women on a scale of preference. The problem, however, was this: how could sexual desirability be quantified? In the early twentieth century, no one could yet imagine a researcher asking women how many men they had intercourse with in their lives, their age at first intercourse, the frequency of their orgasms, the number of men they had engaged in sexual relations with outside of marriage, and the other intimate questions that became standard features of sexual surveys in a later age. The best measures for sexual desirability available to the earlier researchers were rather more indirect: surveys of love poems, questionnaires asking subjects to list the most attractive physical features of the opposite sex, and calculating the frequency of marriage per beauty category and the age of first marriage according to beauty rating. Fertility could be assessed through tabulating the age of first menstruation, the number of pregnancies, and the age of menarche for each female subject.

Early-twentieth-century discussions of women's physical traits perceived as attractive by men were subjected to increasingly rigorous statistical analysis as the decades progressed. Hall's 1907 work on adolescence, using data from Frank Drew's study of Clark University students' love poems, found that men most often found attractive a women's eyes, hair, and face, in that order. They often preferred women with a particular hair color, attractive shape of hands and fingers (e.g., long, clean nails), a slightly upward curving nose, a long neck, prominent eyes, dimples, perhaps freckles, and a clear and flexible voice. "Mode of laughing," was somewhat less important, followed by carriage, gait, gestures, movement or roll of the eyes, and finally pose of the head and shoulders. Some statistical data was also included: 8 percent of males in the study mentioned preferring women with sloping shoulders; 5 percent said that they were particularly attracted to women with long lashes, 4 percent found arched eyebrows especially attractive, and 3 percent found women with cowlicks "charming."[65]

Roswell Johnson was one of the most important, and most enigmatic, of the early evolutionary psychologists interested in quantitative studies. Johnson was born in Buffalo, New York, in 1877. He studied at Brown and then Harvard, where he became a pupil of Charles Davenport, the leading American eugenicist. His devotion to Davenport was remarkable: he called Davenport his "godfather," sending him copies of his grades and discussing his biological interests in frequent letters throughout Davenport's life.[66] Davenport maintained his part in the relationship, repeatedly intervening to help Johnson in his tortuous career.

In 1899, Johnson followed Davenport to the University of Chicago when the latter accepted a teaching appointment there. Johnson spent his summers at the Marine Biological Laboratory and the United States Fish Commission, both at Woods Hole, Massachusetts. During this time,

Johnson published papers on fetal development, marine life, amphibians, reptiles, and plants. Davenport was particularly impressed with Johnson, calling him "a remarkable man in many respects and especially fitted for this kind of work."[67]

Thus far, it would appear that Johnson was a rising star in the biological field. Unfortunately, however, Johnson's father, who owned a hotel in Chicago, encountered difficulty in paying for his son's continued education.[68] Once Johnson completed his bachelor's degree at Chicago, Johnson took a high school teaching position in Sioux Falls, South Dakota, for one year. He then moved on to the University of Wisconsin, and obtained his master's degree there in 1903. It appeared that he would go on to work as a laboratory assistant at Wisconsin while obtaining his doctorate. But then Johnson made a mistake that nearly cost him his career: he took up vigorously advocating Socialism. As a result, he was let go from Wisconsin.[69] The experience was devastating to Johnson. As he pledged to Davenport: "I assure you that never again will I hazard my position by Socialist propaganda."[70] However, it was too late. Charges of Socialist or Communist sympathies would dog Johnson for the rest of his life.

Johnson took up a position in biology and agriculture at Washington State Normal School, Cheney, from 1903 to 1905. In the summer of 1904 he took courses in entomology and plant breeding with the world renowned geneticist Hugo de Vries at the University of California, Berkeley.[71]

Johnson decided that his interest in experimental biology could not be satisfied at Cheney, and so applied to the Carnegie Institute of Washington (CIW) for a research position with Davenport at the Station for Experimental Evolution at Cold Spring Harbor. Specifically, Johnson proposed to the CIW that they provide funds to construct a vivarium in which he could conduct experiments in evolution. As he explained to the CIW:

> Our knowledge of the processes of evolution has been greatly retarded by the lack of experimental investigation. . . . Evolution, above all other things, requires dynamic studies. . . . I remember that Professor Davenport in his course on evolution at Harvard made a strong plea for experimental work.[72]

Although the Institute did not build the vivarium, they did employ Johnson at Cold Spring Harbor from 1905 to 1908. During the final two years of this period, Johnson also studied at Columbia.

Then follows another inexplicable break in Johnson's career. He left Cold Spring Harbor to work as an oil and gas consultant at the Sagamore Oil and Gas Company in Bartlesville, Oklahoma, from 1908 to 1912. During his time in Bartlesville, Johnson began advocating eugenics in such magazines as *Popular Science*. He wrote that criminals and other "inferior men and women" should be "cut off altogether from parenthood" through "compulsory segregation," which is "terminable only by old age or voluntary sterilization . . ." Johnson was also concerned that immigration was bringing "an influx of inferior peoples from southern and central Europe" to the United States. Once again, eugenic measures were called for.[73] He also contributed to the work of the Eugenics Committee of the American Breeders Association at this time.[74] Johnson developed his interest in human sexual selection while in exile in Oklahoma.[75]

By 1912, Johnson's resumé in both oil production and eugenics was apparently solid enough to secure for him a position as a "professor of oil and gas production" and "lecturer in Eugenics" at the University of Pittsburgh. He would remain at Pittsburgh for the next twenty years.[76]

Immediately upon arriving at Pittsburgh, Johnson began investigating human sexual selection. He zealously advised Davenport that the Eugenics Record Office should turn its attention to the subject. Johnson suggested the outline of several investigations that could generate data in support of the idea that men compete to marry beautiful women:

> An investigation of the marriage rate of (1) phi beta kappa as against the rest of the class and (2) of the classes in co-education as against partial (?) co-education to complete co-education and (3) of the best looking sections of graduating classes in comparison with the rest as judged from the photos in college annuals. I believe the special (?) action of sexual selection well worth investigation.[77]

Johnson was not content merely to make recommendations to Davenport. He guided his student, Carrie F. Gilmore, in conducting what appears to be the first study to discover a correlation between female facial beauty and incidence of marriage. Measuring sexual desirability was problematic for Gilmore and the other researchers in evolutionary psychology in this period. Pointed inquires into the sexual lives of the subjects were not yet possible. As would others, Gilmore had to rely on incidence of marriage as her measure of sexual desirability.

In this simple but effective study, Gilmore had impartial male judges rate the attractiveness (on a scale of 1 to 100) of the 1902 female graduates of Southwestern State Normal School, in California, Pennsylvania. The judges accomplished this task by examining the photographs of the graduates. Gilmore than determined which graduates had married by 1912. About 70 percent of women who scored 80 or higher on the attractiveness scale had married, versus none of women who scored 40 or below.[78] Quite obviously, female attractiveness was dramatically correlated with marriage rates; the more attractive the woman, the greater likelihood that she would marry. Over the next forty years, virtually all researchers in the field of human sexual selection would mention the Gilmore study as evidence that beauty was indeed a cause of sexual attraction.[79]

A number of years transpired before more elaborate studies were conducted on preferential mating. In 1929, Johnson's graduate student Josephine Olivia Naly produced a study with the specific intention of verifying the operation of Darwinian sexual selection in humans.[80] Like her professor, Naly was also a member of the American Eugenics Society.[81]

Before engaging in her own study, Naly assisted with the analysis of data on human sexual selection collected by Dr. Anna Rachael Whiting. In the Whiting-Naly study, three judges rated the attractiveness of photographs from the 1916 senior class of Smith College. By 1928, the most attractive quartile of women in the class (of 337) had a marriage rate from 75 to a 100 percent (depending on the woman's attractiveness ranking). The class as a whole had a marriage rate of only 68 percent.[82]

Naly's own study made use of three Smith College class yearbooks, spanning the years 1907 to 1913. For each year of the study, a different set of five male judges was used. They ranked the women on each page of the annuals (nine women per page) according to which they thought "were the most acceptable as mates," from most to least acceptable. They also chose the 25 most "acceptable" women appearing in each yearbook. By 1929, 72 percent of the most attractive women from each page of the 1907 yearbook were married. Similarly, 71 percent of the 25 most attractive women of the entire 1907 yearbook were married. In contrast, only 57 percent of the least attractive women from each page of the yearbook were married by that date. The other data followed much the same trend. Of women graduating in 1910, 70 percent of the most attractive women were married by 1929, whereas only 53 percent of the least attractive women were married. From the 1913 yearbook, 76 percent of the most attractive women were married, versus 57 percent of the least attractive women.[83] Naly averaged the results from all three graduating classes. She found that the most beautiful women had a marriage rate of 72 percent. The least attractive had a marriage rate of only 56 percent.[84]

Each photograph was then given a composite score based on the individual judges' rankings, from 1 to 10. Thus, if all five judges gave the subject a perfect "10," the subject would attain the maximum score of 50. If we transpose the attractiveness rating to a scale of 100, and correlate this with marriage, we find that about 80 percent of women rating a 100 in attractiveness were married by 1929. This figure falls to a 40-percent marriage rate for women rating 0 in attractiveness.[85]

With these data, Naly could legitimately conclude that the attractiveness of the women in her study was positively correlated with their marriage rate.[86] As with Gilmore, Naly's study would be extensively quoted; however, the years in which such studies were considered to be "good science" were quickly drawing to a close.

By far the most sophisticated preferential mating study of the early twentieth century was performed in Italy. Though this may seem surprising, Italian science in the 1920s and 1930s combined several elements that brought issues of human reproduction to the forefront of science there. In turn-of-the-twentieth-century Italy, the social sciences were the most developed of all sciences. Soon after the country's unification in 1860, Italian scholars combined the strong national tradition of excellence in the humanities with the burgeoning interest in the "hard" sciences to emerge as a powerhouse in the social sciences. By the early twentieth century, the Italian social scientists Cesare Lombroso, Alfredo Niceforo, Paolo Mantegazza, Guigliemo Ferrero, and Corrado Gini were accorded international reputations of the highest rank. Largely through Gini's efforts, Italy had perhaps the most sophisticated demographic research institutions in the world.

Benito Mussolini fostered this Italian focus on the development of national demographics and the application of this knowledge to social control. In the late 1920s, Mussolini turned his attention to Italy's demographic policies as part of his efforts to transform Italy into an aggressive world power. Having recently gained dictatorial control over the country, Mussolini employed Corrado Gini, the foremost demographer and eugenicist in Italy, to aid him in formulating a comprehensive population policy for Italy in line with the Fascist program. Gini himself embraced a "Latin" eugenic program that called for population expansion as the surest means to produce relatively large numbers of superior individuals. The goals of both these men were served by policies encouraging Italians to have large, healthy families.[87]

In his famous "Ascension Day" speech of May 26, 1927, Mussolini announced a number of new demographic initiatives. Italians were told to have large families. The government would provide health care, loans, prizes, and other perks for those who cooperated. Those who didn't would face higher taxes and difficulties in career promotion. In line with the coming rapprochement with the Catholic Church, divorce, birth control, abortion, and eugenic sterilization were officially condemned.[88]

Mussolini also intended to mobilize Italian science in the quest for more and better births. He thus ordered the formation of an Italian Commission for the Study of Population, and appointed Gini as its president. Gini was authorized to conduct whatever research was necessary to support the state's population goals, and to present the Commission's research to the International Congress for the Study of Population that would meet in Rome in 1931.

Gini, with the help of the internationally renowned endocrinologist Nicola Pende and a squad of research assistants, fanned out throughout Italy collecting data on the physical characteristics of men and women that seemed to be correlated with high levels of fertility.[89] Gini also employed his own research assistant in the study, Carlalberto Grillenzoni. Gini asked Grillenzoni to investigate the possible linkage of physical attractiveness, body type, fecundity, and marriage age in women. In a particularly prescient move, Gini also wanted Grillenzoni to determine if attractive women also chose attractive men as spouses.

Grillenzoni's study was based on an analysis of 1,500 women between thirty and sixty years of age.[90] Specifically, the study measured the women's elegance of dress; physical beauty; body type; height; cosmetic appearance; and physical appearance of their husbands (if married), and correlated this data with the women's marital status; age at marriage; number of children born; and length of marital fertility (the length of time between marriage and the last child). The study also compared the attractiveness of the spouses.[91]

Grillenzoni found that whereas 75.6 percent of the least attractive women married, 93.6 percent of the most attractive women married.[92] It should be noted that a comparison of Grillenzoni's study with the Udry-Eckland study discussed in chapter 2, which was conducted fifty-two years later, shows an amazing similarity between the two studies in their hypotheses, methods, results, and conclusions.

Grillenzoni also compared marital status to degree of attractiveness. He found that unmarried women were less attractive than married women. While unmarried women rated 1.27 on his scale of attractiveness, married women rated 1.64.[93]

The age at which the women contracted their first marriage was correlated with their beauty rating. The least attractive women were first married at age 26, on average. The most attractive women, however, married younger on average, at 22.7 years old.[94] Once again, Grillenzoni's study used essentially the same methodology to arrive at the same conclusions as the much later Udry-Eckland study. Reversing the order of variables, Grillenzoni then correlated the age of first marriage with average degree of attractiveness. He found that the youngest brides, who married at less than nineteen years of age, had an average degree of attractiveness of 1.84. Brides older than thirty-five at their first marriage were less attractive, with a score of 1.27.[95] Finally, Grillenzoni expected that the women who married youngest might contain an unusually high proportion of attractive women. He found that 5.8 percent of the brides married under age eighteen received the highest attractiveness rating, whereas only 1 percent of the youngest brides received the least attractiveness rating.[96]

Grillenzoni concluded that the relationship between marriage and beauty was in accordance with the earlier studies: "It is therefore indisputable that beauty is an important variable in matrimonial selection. . . . Between beauty and the probability of contracting a marriage or the precocity of marriage there is a notable positive relation . . . we can easily understand how beauty would make marriage more probable and more precocious . . ."[97]

Grillenzoni next set out to examine the relationship of body type with marriage and childbirth. As Devendra Singh and others would also conjecture half a century later, Grillenzoni hypothesized that women with regular figures would not only prove to be more sexually attractive, but also be more fertile. This, in fact, proved to be the case. Whereas 8.2 percent of women with a "thin" body type remained unmarried, and 9.4 percent of women with a "regular" body type did not marry, over 17 percent of women with a "heavy" body type remained unwed.[98] Body type was then paired with average age of first marriage. Women with a regular body type married youngest on average, at 24.27 years old. Women who were thin had an average age of first marriage of 24.57 years. Heavy women on average were 25.34 years old when they first married.[99] Once again, the results were clear: regular figured women were seen to be more desirable as mates. Heavy women were especially at a disadvantage in this respect.[100]

Next, Grillenzoni examined body type and average number of children born to each category of women, be they either married or unmarried. He found that married women with a regular figure bore an average of 3 children. Thin married women had 2.55 children. Heavy married women had only 2.41 children.[101] Also, Grillenzoni discovered that women with no children or with only one child were more often thin or heavy than regular in weight. Finally for this segment of his study, Grillenzoni calculated the average length of time from marriage to the last child's birth, according to body type. For women with a regular figure, their average length of matrimonial fertile period totaled 7.84 years. For thin women, this period lasted on average 6.86 years. For heavy women, it lasted only 6.31 years.[102] Grillenzoni's conclusions matched those of later evolutionary psychologists. Not only were regular-figured women more attractive, but they were also more fertile. Grillenzoni announced: "The difference is beyond any doubt . . . the women best suited to reproduction are those whose figures deviate the least from normal. *In medio stat virtus.* . . . Every deviation from the regular figure for women unfavorably influences prolificacy and age of marriage," and always more so for heavy women than for thin women.[103]

Grillenzoni concluded his study with an interesting twist unique in the annals of early evolutionary psychology. Unlike the earlier studies,

Grillenzoni sought to determine if there was a significant statistical corre-lation between the attractiveness of the spouses. He determined that this, indeed, was the case. The least attractive women married husbands whose attractiveness rating was only 1.19. The most attractive women had more attractive husbands. They rated 1.85 on Grillenzoni's scale.[104]

Mussolini was undoubtedly pleased with the results Gini and his researchers obtained. Mussolini had claimed that the fashionable "boyish figure" of the 1920s was antithetical to Fascist "family values," and the results of the Population Study seemed to confirm this. He now ordered Italy's medical establishment to combat "the fashion of excessive thinning down." Italian men were now expected to find buxom, fertile women beau-tiful. The anti-Fascist historian Gaetano Salvemini would snidely refer to this program as "the battle for fat."[105]

Gini was aware that Grillenzoni's study was the most important that had yet been accomplished in human sexual selection. As Gini wrote in the introduction to Grillenzoni's study, "it is hoped that the suggestive results of Miss Gilmore and Miss Naly and the most important findings in this work by Dr. Grillenzoni will induce a greater and more systematic research into a topic which merits more attention than it has so far received, given its great biological and social importance."[106] However, Gini's pious wish for further study would go virtually unheeded. And this may not have sur-prised him. As an internationally known and respected scientist and a leading eugenicist, in constant contact with his colleagues in America and elsewhere, Gini may well have suspected that the tide was turning against such studies. He became rather irritated six years later, after seeing Samuel Holmes's rather more limited study on the relationship of beauty to scholastic achievement and marriage selection in the journal *Human Biology*. Gini wrote a note to the journal's readers reminding them of the Gilmore, Naly, and Grillenzoni studies on human sexual selection that had been completed years earlier.[107]

Samuel Jackson Holmes was another pioneer in human sexual selection studies. Holmes originally wanted to enter the medical profession, but when he arrived for study at U.C. Berkeley he quickly found himself attracted to marine zoology. In 1894 he helped explore the California coast in an effort to find a suitable location for a seashore laboratory for the University. While on this expedition, Holmes discovered several new species of *Crustacea*, a class he would continue to study in later years. Holmes went to the University of Chicago for his Ph.D. in 1897, and taught at Michigan and Wisconsin for a number of years thereafter. In 1912 he accepted a professorship in zoology at Berkeley, where he would remain for the rest of his career. His interests turned more toward genetics and eugen-ics in the succeeding years. His most notable achievement was the discovery that radiation induced genetic mutations. The *New York Times* would eventually proclaim Holmes as "one of the world's foremost authorities on zoology and genetics."[108] Holmes was particularly interested in the practical application of his sciences to society: he became a rather prolific researcher and writer on evolutionary psychology, eugenics, and medical ethics.[109]

Holmes's most important research on human sexual selection came toward the end of his career, in the late 1930s. In 1938 he published with C.E. Hatch the last study on human sexual selection that would be performed for over forty years. In "Personal Appearance as Related to Scholastic Records and Marriage Selection in College Women" they examined the correlation between female beauty, college GPA, and marriage. They justified their study by what was now an old refrain: "According to Darwin's theory of sexual selection, beauty has played an important role in selective mating both in human beings and in the higher animals below man. . . . it seems probable that the selection of mates on the basis of aesthetically pleasing appearance has contributed somehow to the biological welfare of the species."[110]

Their study utilized three male judges to rate the beauty of 642 women at Berkeley. The women's appearance was ranked using four categories: I beautiful; II good looking; III plain; IV homely. Judges' scores for beauty were the same in 95 percent of cases, leading Holmes and Hatch to conclude that "beauty represents something about which the judgments of different people show a fair amount of agreement."[111]

The women's attractiveness assessment was then compared with their marriage rates several years after graduating from college. Whereas 34 percent of beautiful women were already married, and 28.05 of "good looking" women were married, only 15.66 percent of "plain" women were married and just 11.25 of "homely" women were wed.[112] Consistent with the earlier findings, Holmes and Hatch concluded that "clearly . . . the percentages of women graduates who marry within a few years after leaving college regularly decrease as the ratings for beauty become lower . . . thus affording evidence that beauty plays a very important role in marriage selection within the class of individuals studied."[113]

Seventeen years earlier, during the height of interest in sexual selection, Holmes had shown an interest in how the characteristics that attracted men and women to each other differed between the sexes. He realized that this would go beyond mere beauty, to include such ineffable traits as intelligence, temperament, general physical health, and so on. In the expected "assortative preferential selection," one sex might "trade" some characteristic they excelled in for a different trait sought in the other. For example, women might "trade" their beauty, desired by men, for a man's supporting resources. At the same time, men might seek to attract beautiful women through advertising their command of resources or status. This hypothesis could be tested by examining a ranked listing of desired characteristics in a mate, according to sex. Holmes discovered that such a study had, indeed, been performed by *Physical Culture* in 1915.[114] Among other findings (to be discussed later in this chapter), the study confirmed the high value men placed on healthy, good-looking wives.[115] In descending order of importance, the male respondents rated health as most important (23 percent), followed by "looks" (14 percent), housekeeping skills, disposition, maternal character, education, "management" (presumably of the household), dress, and character.[116]

Eugenicists were pleased that those traits they most admired, such as health, beauty, hard work, and interest in family, seemed to top out on this and other charts. Leonard Darwin learned of this study through Paul Popenoe and Roswell Johnson's text, *Applied Eugenics*. Darwin thought Holmes's work was important enough to merit mentioning it in his address to the Eugenics Education Society, in 1922:

> In America, young people having been asked to state what were their ideals in connection with matrimony, the results being at first sight on the whole very satisfactory. Both sexes placed health at the top of the list of the desirable attributes of a prospective mate, whilst disposition and education were also rated highly. If good looks appeared to count for a good deal, it must be remembered that pink cheeks are correlated with a good constitution, with healthy habits, and with youth, this last being of great importance from a biological point of view.[117]

The 1920s ushered in a veritable stampede of professors seeking to determine the traits men desired in their spouses, through questioning their college students. Harrison R. Hunt, a University of Mississippi zoology professor, published such a study in the *Journal of Heredity*, one of the most prominent eugenics-oriented journals of the time. Hunt feared that eugenic values were not sufficiently guiding young college graduates in choosing their marriage partners. To determine if this was the case, Hunt sent a questionnaire to 555 students of his university asking them to rank in order of importance those traits they felt were most important in a potential spouse.[118]

Fortunately for Hunt, his fears were allayed. His results showed that men and women put significant emphasis on healthy, intelligent, and morally sound partners. There were the "expected" gender differences in traits desired: "The men rate the willingness to have a family and physical attractiveness significantly higher than do the women. . . . Conversely the women rate the following traits significantly higher than the men do: ambition, business ability, and mutual intellectual interests. . . . the women emphasize more than the men the importance of business sagacity and the determination to succeed."[119] Hunt concluded that "For the most part the attitude of the students in the University of Mississippi is morally and eugenically good."[120]

Hunt ended his paper with his own take on the nature–nurture controversy: "The importance of environment in providing the fertile soil in which the seeds of hereditary promise may grow, should not be minimized. But the prevalence in democracies like ours of the fallacious dogma of human equality makes it imperative to emphasize the fundamental importance of heredity."[121]

Hunt's interest in eugenics only grew when he left Mississippi for a position at Michigan State University. There he was active in the Eugenics Research Association and the International Federation of Eugenics Organizations. His research interests focused on the eugenic affects of war and on twin studies. Some of this research was funded by Charles Matthias Goethe, a California-based millionaire banker and eugenics benefactor, and published by the Galton Society.[122]

Similar studies by others followed. In a study conducted by Edwin L. Clark at Ohio State University, the male students desired wives who were of the same race as themselves; from "sound family stock"; healthy; physically attractive; intelligent; affectionate; and interested in having children. "Good housekeeping and the ability to care for and train children were also regarded as necessary."[123]

At New York University, Rudolph M. Binder offered his male students only a very limited selection of traits (health, wealth, or beauty), from which they were asked to select the trait most important to them in a future wife. Forty-four percent of the men ranked health as their chief concern in choosing a spouse; 42 percent selected beauty. Only 26 percent wished to choose a wife for her wealth.[124] Compared to the results from his female students (see below), a considerably greater percentage of men chose beauty as the most desired trait.

Traits Women Sought in Men

Sexual selection models presumed that both sexes exercised choice in choosing mates, though these choices would vary depending on the reproductive needs or advantages of each sex. Critics, however, conceived of a serious obstacle in the application of this hypothesis: perhaps females never had the luxury of choice, due to social norms that might allow only men to choose mates. Furthermore, the greater physical strength of the male might be utilized to savagely prohibit women from choosing with whom they would

mate. If the cartoonish notion of the cave man clubbing a woman unconscious and dragging her back to his cave prevailed, it would constitute a
major defeat for the evolutionary psychologists. Even such a staunch
evolutionary psychologist as Gerrit Miller Jr. wondered if not countervailing
behaviors such as rape posed a threat to female sexual freedom.

Samuel Holmes concerned himself with defeating this objection. He
assembled evidence from various authorities acquainted with the practices
of severely patriarchical societies in support of the "women's choice"
model. Holmes found proof in G.E. Howard's *History of Matrimonial
Institutions* that even in societies that practice wife-purchase, women have
influence over their marital fate. Using Howard's observations, as well as
the findings of Westermarck and other authorities, Holmes affirmed that
women's "liberty of selection . . . is very considerable, and however down-
trodden, they well know how to make their influence felt."[125]

Next, evolutionary psychologists had to determine what characteristics
women found attractive, and why. A woman naturally sought to marry a
man who could protect her and her children, and provide for their material needs. She also wanted a partner who focused his attention on her and
her children, sought to amuse her, and would not abandon her for another
sexual partner (the tension between men's and women's sexual goals is
obvious here).

Given the ubiquitous sexism of the era, male scientists were naturally
attracted to demonstrating that the biologically frail woman needed a
strong man by her side. John Nisbet emphasized the desire women showed
for a strong man. He believed that this instinct was so deeply ingrained in
the human psyche that "few men, however intellectual, would refrain from
showing off their physical prowess before ladies if the opportunity arose."
Furthermore, not even the most intellectual woman could deny "the force
of such a bid for her admiration," Nisbet added.[126]

The Italian social scientist Guigliemo Ferrero believed that the division
of labor between the sexes, with women nurturing the young while men
protected the family and gathered the resources it needed, was a fundamental consequence of evolution. In an article discussing the gender division of labor in humans, Ferrero briefly considered the responsibilities of
each sex in *Hymenoptera*. He saw the primary duty of the male as impregnating and defending the female and her offspring. Once this task was
fulfilled, the male's role in the cycle of life was complete and he died. Only
in those species in which the male spent his life caring for his family did the
male live a relatively long life. In humans, women will select as husbands
those men who can best fulfill this role.[127]

Edward Westermarck also took up the subject. The evidence he gathered showed that women in all societies tend to select among rival suitors

the man who was most muscular, courageous, and skillful. "We may assume that women's instinctive appreciation of manly strength and valour is due to natural selection in more than one way. A strong and courageous man is not only a likely father of strong and courageous children, but he is also better able than a weak and cowardly man to protect his offspring."[128]

Samuel Holmes imagined that women, millennia ago, sought the "successful warriors," men who had earned the approbation of their tribe. They would most likely be men of greater size and strength, broad shouldered, and aggressive. However, women would have also sought out men with a softer side.[129] Their ideal mate would also be cheerful, vivacious, and of course good-looking.[130] Roswell Johnson reminded his readers that women would also search, of course, for a "steady provider."[131] Knight Dunlap added height to the characteristics of the ideal man.[132] The desire for a tall spouse was instinctive, Dunlap believed. In prehistoric times, a woman vulnerable during pregnancy or while nursing desired the added protection a tall man afforded against predatory animals or other men. Furthermore, height gave an advantage when hunting or competing against other men for resources.[133]

With the Great War still fresh in mind, Dunlap made an interesting observation: women at that time seemingly had been attracted to men displaying military ornamentation. Perhaps, Dunlap suggests, this attraction harks back to the fascination with grand male sexual display shared by both women and female animals.[134] Gallichan suggested that male energy and vigor, rather than grace, were most attractive to women.[135]

The Hunt study of 1921 and the Buss and Barnes study of 1987 both suggest that women place a slightly higher value on marrying an intelligent man than visa versa. John Nisbet would not have been surprised at such findings. He believed that in the course of human evolution, women acquired an instinctive attraction to men's intelligence, as well as to their good looks and strength. In 1890, Nisbet wrote that a woman "instinctively feels that in the battle of life as it has now to be waged her ugly but intellectual friend will prevail over a brainless Adonis. That fine physique is still an element of beauty it would be idle to deny; it is no longer, however, the only one."[136] The 1916 *Physical Culture* study of the ideal husband (discussed at length below) found that native intelligence was more important to women, when imagining an ideal husband, than was his degree of education.[137]

Though evolutionary psychologists discussed women's mating strategies from time to time, they were less concerned with women's mating desires than with men's. There are several possible reasons for this. For one, their patriarchical society felt most comfortable considering men's sexual aggressiveness as opposed to women's. Also, since most of the leading evolutionary psychologists were men, or in a few cases were female students under the supervision of male professors, the focus on male sexuality might simply have reflected a male fascination with their own sexual behavior.

As a result of this bias in sexual interests, few quantitative studies on female sexual preferences were produced in the early period. G. Stanley Hall found in his survey of adolescent sexuality that girls fantasized about romance with well-dressed boys with broad shoulders and white teeth. They disliked boys with deep set eyelids, full necks, ears that stood out, eyebrows that met or were bushy, long feet, high cheek bones, light eyes, large noses, long necks or teeth, or short stature.[138]

In a somewhat more serious vein, Paul Popenoe and Roswell Johnson publicized the results from a spousal preference study examining women's visions of an ideal husband conducted in 1916 by *Physical Culture*.[139] There was no surprise here: women preferred men who were notably healthy, good looking, successful, kindly, intelligent, and enjoyed parenting.[140] Of course, studies conducted seventy years later would obtain essentially the same results.

In Edwin L. Clark's Ohio State University study, women responded that a "good provider" was much more important to them than wealth or fame. Women desired husbands who were sincere, honest, fair-minded, truthful, "wholesome in thought and action," affectionate, intelligent, companionable, and desirous of children who they would love. Physically, these ideal husbands would be of the same race as their wives, come from good families, be healthy, and be handsome.[141] It's interesting to note that, whereas men in this study spoke mainly of the desired physical attributes of their ideal wives, the women stressed the psychological attributes desired of their future husbands.

Robert T. Hance found that freshmen women at the University of Pennsylvania had marital desires similar to their sisters to the west. In particular, the women mentioned the importance of a prospective husband's "frankness" and his ability to have healthy children. (One wonders if "frankness" here was a code word for "marital fidelity.")

As in Harrison Hunt's study, Rudolph Binder found that nearly two-thirds of his female students valued a husband's health above all. Good looks ran a very distant second (20 percent), whereas wealth finished last (16 percent). Binder also noted that quite a few female respondents laid "considerable stress" on their wish for a husband who exhibited cleanliness (perhaps reinforcing the desire for health).[142] In general, all of the results from this multitude of college studies were consistent with Darwinian sexual selection theory.

Sexual Selection,
Male versus Female Reproductive Strategies,
and Family Structure

Although the concept would be much further developed by Robert Trivers in 1971, the early evolutionary psychologists were aware that female and male mating strategies differed. Evolution would favor women who tended to be "choosy" in selecting their sexual partners. Since a woman would bear only a few children who would live, even after devoting a substantial amount of resources to each child, she would have relatively few opportunities to choose a father for each child. Men chosen on the basis of health, strength, intelligence, and command of resources would likely father healthy children who might have a better opportunity of surviving in a harsh world.

Men, on the other hand, would likely seek larger numbers of mates throughout their lifetimes because their reproductive success would generally be improved by attempting to impregnate relatively large numbers of women. Women's strategies for selecting strong and resource-rich men would also foster the evolution of male aggressiveness. Thus, men would have to compete among themselves for the most desirable mates. In essence: women sought to attract strong, healthy, and resourceful men by displaying their beauty and healthfulness; men then fought each other (either directly or indirectly) to mate with the most desirable women.[143] The least desirable men would most likely not mate at all.[144]

Men's Mating Strategies

In the early twentieth century, a spirited debate developed over the relationship between male sexuality, family form, and evolution. Was the male sex drive sufficiently strong to overpower any cultural or social inhibitions that promoted monogamy? Most evolutionary psychologists maintained that it was. Charles Darwin believed that, if so, the strongest and most socially powerful men would acquire the most wives, thus increasing their chances of leaving numerous children.[145]

Gerrit Miller Jr. was the most authoritative voice in favor of this hypothesis. Miller stepped into the debate in the late 1920s, apparently exasperated at the increasingly pervasive assertions that human behavior was unique in the animal world. Miller was convinced otherwise. He argued that all primate behavior with which he was familiar, including the human examples, pointed to only one conclusion: the human male was instinctively promiscuous, as were his simian cousins. Miller proved his point through several lines of evidence. He saw human taboos as socially mandated inhibitions on the instinctual sexual behavior shared by humans and the other primates. As he explained, "taboos are largely directed against behavior which would not be questioned among monkeys . . ." Furthermore, "most of the infractions of the conventional code fit into the picture of generalized primate behavior which has been outlined in the laboratory. It is therefore perhaps not unfair to recognize these infractions as, fundamentally, instinctive returns to the way of the promiscuous band . . ."[146]

The polygamous theory would also explain the persistence of prostitution in human societies. In this case, we should consider prostitution "not as a social phenomenon generated by the forces of specialized civilizations, but as the commercialized survival of a promiscuous condition through which every human society has probably been forced to pass, by virtue of man's primate sexual behavior type." Comparative psychology, the child of evolutionary theory, thus explained why humans tended toward engaging in illicit sexual behavior, a mystery that had puzzled moralists since time immemorial.[147]

Women's Mating Strategies

The nature of women's mating strategies also received some attention. In 1898, Karl Groos recognized that women would employ "coyness" in their quest to "preserve their strength" for their offspring and to insure that they select the best fathers for their children.[148] William McDougall, in the 1919 and later editions of his 1908 masterpiece *Introduction to Social Psychology*, mentions that the females of many animal species, including humans, employed both sexual self-display and coyness as strategies to attract mates. Coyness necessitated the "active pursuit and courtship" of the female by the male, and allowed the female to be selective in which males were permitted to copulate with her. Without such female choosiness, a superior male's strength, skill, "beauty of voice or form or colour" would be much less significant in affecting evolution. McDougall concluded: "The probability that female coyness plays this important role in evolution affords some ground for the view that it is the expression of a special instinct whose function it is to give scope for sexual selection."[149]

The use of the term "coy" by Groos and McDougall to describe female caution in selecting a mate was prescient. Late-twentieth-century evolutionary psychologists would employ the same term to describe the caution women use in withholding coitus while they assess a potential mate's reproductive value.[150]

Nevertheless, the desire to use various means to attract a suitable mate prevails in women. In the *Descent of Man*, Darwin noted: "women are everywhere conscious of the value of their own beauty." He attributed this to their desire to catch the attention of potential mates.[151] G. Stanley Hall concurred. He realized that women's selectivity, in this manner, played a critical role in guiding male evolution. For example, Hall surmised that the aggressive instinct of men had been mitigated over time in favor of a more "esthetic" approach to sexuality. Women, he wrote, have been "a constant biotonic stimulus" on the evolution of the male psyche.[152] Many eugenicists, such as Roswell Johnson, welcomed the effect that the female proclivity to select the "best man" had on guiding human evolution in a favorable direction.[153]

Anne Campbell and other evolutionary psychologists predict that, like males, females will compete with each other to display their attributes most likely to attract the opposite sex.[154] They will tend to exhibit jealousy and even hatred for those rivals who they fear "beat" them in this competition. Hall would have emphatically agreed. He observed that puberty brought with it the advent of sexual rivalry. Animals would compete for the attention of the opposite sex through "animal battles" and "showing off." Similarly, adolescent girls often displayed intense rivalries related to sexual competition. "It is hard for girls to admit that others are more beautiful, witty or cultured than themselves, and rivalry often drives them to extreme or even desperate acts." Cesare Lombroso and Guigliemo Ferrero believed it inevitable that sexual rivalry would eventually destroy the bonds of friendship between two adolescent girls.[155]

Women were also attuned to the threat that their spouses might seem attractive to other women, who could then "steal their husbands away." This anxiety reached the point that some women disdained the marital advances of men they judged too attractive, for fear that they would suffer in a marriage where the husband frequently provoked his wife's jealous suspicions.[156]

Sexual Competition and Jealousy in Males

The concept of "competition" was particularly attractive to scientists of the pre–World War II years. Everything from Darwin's writings to class and national rivalries seemed to suggest that competition for scarce resources ranked next to the law of gravity as a fundamental constant in the universe. Sociological and anthropological studies conducted in the early twentieth century certainly conformed to these expectations.

Sexual competitiveness manifested itself through jealousy. G. Stanley Hall, in his investigations into adolescent psychology, could not help but notice the eagerness of teenage boys to display their physical prowess whenever girls were present. Hall described this behavior, present in animals as well as in humans, as the "showing off" instinct. His massive survey of adolescent behavior confirmed this observation and its implication for sexual competition:

> Hundreds of boys, in our returns, run fastest, hit hardest, talk loudest, are most stimulated to compete and excel, do rash and foolhardy or unusual things, when observed by girls, or perhaps one in particular. . . . Older youth are not without sex consciousness in the display of athletic feats in which the body is more or less exposed.[157]

Hall explained this phenomenon in anthropological terms:

> In a primitive polygamous state, where each male desires as many females as possible, he is at war with all other males and frequently in a life-and-death struggle with them. He often wars on neighboring races for the capture of wives, where exogamy is the custom. Where the female is the prize, victory may be defined as successful courtship and war is for the sake of love.

As civilization developed, actual fighting was replaced with more symbolic contests.

Edward Westermarck agreed with Hall. Westermarck added that in numerous human groups, such as the Guanas of Paraguay, the Eskimo of the Bering Strait, the indigenous Australians and other peoples, men customarily engaged in combat for desirable women in accordance with their particular tribal customs. Similar competition for mates was observed in certain species of birds, moths, spiders, and pheasants, Westermarck noted.[158] Westermarck realized that male jealousy was stimulated by a man's fear that another man's offspring might be born into his family unawares.[159]

In his discussion of jealousy, Havelock Ellis offered the explanation of fellow eugenicist Arnold L. Gesell for the cause of this emotion: "Viewed broadly, jealousy seems such a necessary psychological accompaniment to biological behavior, amidst competitive struggle, that one is tempted to consider it genetically among the oldest of the emotions, synonymous almost with the will to live, and to make it scarcely less fundamental than fear or anger."[160]

William McDougall also saw a connection between sex and the instinct of "pugnacity" in males:

> The assumption of a specially intimate innate connection between the instincts of reproduction and of pugnacity will account for the fact that the anger of the male, both in the human and in most animal species, is so readily aroused in an intense degree by any threat of opposition to the operation of the sexual impulse; and perhaps the great strength of the sexual impulse sufficiently accounts for it.[161]

McDougall then offered a fascinating account of the interplay of jealousy, social structure, and evolution. Based on speculations regarding Paleolithic family structures by J.J. Atkinson and Andrew Lang,[162] McDougall imagined that prehistoric families were polygamous entities composed of a patriarch and his wives and children. The patriarch, keen to prevent any challenges to his dominance over the women of the community, drove out the young males as they reached adulthood. These exiles formed semi-independent bands, which continued to be loosely associated with their paternal community. Occasionally sexual desire would impel some of the young males to compete with the patriarch for a nubile wife. Once one of the challengers succeeded in displacing the patriarch, the victor would become the new patriarch while continuing the age-old pattern of "fierce sexual jealousy." This pattern, McDougall noted, obtains in some animal species, and would appear to apply to humans as well, were legal and moral restraints removed. Indeed, McDougall believed that the patriarchal restraint of younger men's sexual urges was the first law, so to speak. This law was not put into place for rational reasons, however, but because a stronger male was able to employ force to satisfy his sexual desires in the face of other men whose own desires were thwarted. The willingness to use force in this sexual combat was itself instinctive, McDougall asserted. Assuming that strength, ferocity, and physical attributes were assets in such a struggle, these battles over sexual mates likely influenced human evolution. Given that the challenger risked death if he failed, those who were reckless but weak would be eliminated from the population. On the other hand, those who were too fearful to fight for a mate certainly would not reproduce, and so the timorous were also unable to pass on their characteristics to their offspring. The balance of these forces tended to encourage the propagation of those men in whom prudence, rather than recklessness or fear, was the dominant characteristic. Prudence is a sophisticated mental state, implying "a considerable degree of development of self-consciousness and of the self-regarding sentiment and a capacity for deliberation and the weighing of motives in the light of self-consciousness." Thus, the struggle for mates would tend to encourage the evolution of a strong sex drive and physical aggressiveness, but also a thoughtful prudence. Men with these characteristics would likely become succeeding patriarchs, and transmit these traits to their offspring. The traits of self-control and "law-abidingness" thus engendered would form an essential backdrop to the development of civilization.[163] The reader may note the tantalizing similarity between McDougall's account of the influence of jealousy upon human evolution and more recent discussions of the same topic.

Sexual competition seems even to have evolved an inter-cellular attack mechanism in humans. Edward Westermarck, in 1922, suggested that sperm of different males might "counteract" each other in the female birth canal. Seventy-three years later, Robin Baker and Mark Bellis confirmed Westermarck's suspicion, calling the variety of spermatozoa imbued with this attack property "killer sperm."[164]

*   *   *

The evolutionary psychologists' work on human sexual had thoroughly considered the biological motivations of sexual behavior, and demonstrated its hereditary nature using two basic lines of evidence. Human sexual behavior was analogous to sexual behavior found elsewhere in the animal world. The closer an animal was to humans in evolutionary terms, the greater was the similarity in their sexual behavior. Furthermore, several studies presented hard statistical evidence that human sexual behavior conformed to expectations predicted by the theoretical model.

Success in demonstrating the powerful influence of heredity on sexual behavior encouraged the examination of other areas of animal behavior that were still poorly understood. Even a causal glance at nature would suggest that humans shared a remarkable range of behaviors with other animals, including the formation of family units, acts of self-sacrifice, and acts of great cruelty. Some evolutionary psychologists proposed that there was a hereditary explanation for these, as well.

5

Evolution, Ethics, and Culture

Cruelty, selfishness, lust, cowardice and deceit are normal ingredients of
human nature which have their useful role in the struggle for existence.
Intrinsically they are all virtues.

Samuel J. Holmes, "Darwinian Ethics," p. 123

It is . . . altogether erroneous to suppose that human institutions, such as
the state, religion, marriage, etc., are purely arbitrary products somewhere and at some time by chance devised by a ruler or a ruling group
for their own ease or their own advantage. Were this so, were these institutions not founded upon the inner life of man as a whole, with all its
instincts and impulses, they would long ago have disappeared like a passing craze of fashion, and fallen into oblivion.

Friedrich Alverdes, Social Life in the Animal World, *p. 198*

Altruism

Like modern evolutionary psychologists, their predecessors concerned
themselves with understanding the evolutionary significance, and perhaps
even origin, of ethical behavior. Indeed, this was quite provocative. It was
widely assumed, in both eras, that even if human beings retained some
instinctual aspects to their behavior, certainly their higher ethical norms
and ideals were the products of millennia of cultural progress. To question
this by asserting a hereditary, biological foundation for human ethical
conduct was to seemingly question the very notion of "culture" as defined
by the environmental behaviorists. Undoubtedly evolutionary psychologists were on softer ground here than in their more readily observable
claims that instinct influenced human sexual selection. Nevertheless, they
accepted the challenge, and strove with some success in creating the

theoretical foundation and finding empirical evidence to support the contention that at least some of the core elements of human ethical behavior were derived from instinct. In the process, they would anticipate some of the most important ideas developed much later by William Hamilton, Edward O. Wilson, and Randy Thornhill regarding altruism, kinship selection, and the possible causes of rape.

Certainly the apparent existence of altruistic behavior in humans and animals was one of the more interesting puzzles facing evolutionary psychologists. Why did humans, and even some animals, engage in what appeared to be altruistic behavior? Wasn't this in direct opposition to Darwin's idea of "survival of the fittest," essential to contemporary principles of evolution? Beginning with Darwin's analysis, early evolutionary psychologists did develop plausible hypotheses to explain altruism in its evolutionary context, and found evidence to support their ideas, primarily through primate studies.

Darwin confronted altruism as one of the most obvious challenges to his theory. His explanation had to include some aspect of hereditary behavior that increased the likelihood of survival of the organism, in exchange for devoting resources to others. Darwin's response rested on the assumption that if each member of a social group aided the others, the survival of all would be enhanced to a greater extent than if the organism engaged in only purely selfish behavior. Thus, natural selection would favor group altruism. Darwin believed that altruism evolved because:

> the social instincts lead an animal to take pleasure in the society of its fellows . . . and to perform various services for them. The services may be of a definite and evidently instinctive nature; or there may be only a wish and readiness, as with most of the higher social animals, to aid their fellows in certain general ways . . . those communities which included the greatest number of the most sympathetic members would flourish best, and rear the greatest number of offspring.[1]

Before Hamilton's kinship selection theory was developed in the 1960s to explain altruism, most sociobiologists and evolutionary psychologists sought to validate Darwin's idea that altruism evolved primarily as a group protective mechanism. Karl Pearson, William Trotter, and Samuel Holmes all accepted Darwin's explanation for group selection. Karl Pearson stressed that strengthening the instinct for "group cohesiveness" would almost certainly increase the survival of a group and its offspring.[2] As an example of the benefits of group solidarity, William Trotter noted that in carnivores, the herd is stronger than the individual, better able to hunt, more sensitive to its environment, and more responsive to impending danger (due to the perceptions of alarm of other members of the group) than is the individual. Furthermore, Trotter claimed, the herd makes more efficient use of its food resources (e.g., eating all of a kill).[3] Warder Clyde Allee, among others, pointed out that the enhanced likelihood of group survivability would also explain human social altruism.[4]

Comparative psychologists eagerly sought to find evidence for an altruistic instinct in chimpanzees. The assumption was that chimpanzee altruism would be similar enough to that of other higher mammals, and basic enough in its manifestation, that its instinctive origin would be obvious. At the same time, the altruistic behavior would hopefully be complex enough, and sufficiently similar to analogous human acts, that the link to human altruism would be unmistakable.

In his studies of chimpanzee behavior, Frederich Alverdes mentions some anecdotal evidence of altruism. For example, he had witnessed chimpanzees passing pieces of fruit to the mouths of their friends. "The origin of human kisses most probably lies in this custom," Alverdes speculated.[5]

Robert Yerkes and his associates took up the challenge of proving the existence of chimpanzee altruism. Indeed, this was one of Yerkes's principle goals. Chimpanzee altruism would demonstrate that many human behaviors were influenced by hereditary instinct. In his article, "Mental Evolution in the Primates," Yerkes asserted that "through the several classes of primate one may trace the evolution of social consciousness and its behavioral manifestations toward mutuality of interest, self-subordination, cooperation and altruism."[6] After years of observation, in 1933 Yerkes announced that he could now finally demonstrate the existence of chimpanzee altruism with an obvious connection to human behavior.

Over time, Yerkes had observed numerous instances of an amazingly developed sense of individual recognition, ability to interpret behavior, and capacity for reciprocal exchange in chimpanzees. Chimpanzee reciprocity was based on the individual's confidence in another individual, he concluded. Confidence, in turn, was founded upon an intuitive appreciation of trustworthiness, fairness of treatment, justice, and honesty. The similarity of Yerkes's observations to those of Robert Trivers and others on animal and human altruism some forty or more years later is striking.[7] As researchers would again observe in the 1990s, chimpanzees knew when they were being cheated or harmed by another, and would eventually retaliate. Yerkes observed:

> To treat a chimpanzee deceitfully or unjustly and to be detected in such unfairness is tantamount to inviting trouble, for the mistreated individual is likely to seek retaliation or revenge even after long delay. Experience indicates that no man is thereafter safe with a chimpanzee which he has humiliated by ridicule (laughed at boisterously and, as viewed by the animal,

without cause), wantonly injured physically, disagreeably tricked, or otherwise treated with unfairness or unkindness.[8]

Chimpanzees who "liked" one another would engage in grooming each other periodically and in other altruistic acts. This act was healthful for the chimpanzee being groomed, but absorbed the time and attention of the groomer. However, the groomer would expect that, in turn, the initial beneficiary eventually would groom him. Thus, this apparently altruistic behavior was actually self-serving. Yerkes ensured that the reader could not escape the implications of his findings:

> The observations which are summarized in the above general statements clearly bear importantly on the contention herein advanced that grooming in chimpanzees is a form of social service, with mutuality of interest and feeling and with definite altruistic quality. In them, as in us, discovery of willingness to be of service, more or less disinterestedly, encourages confidence. Extreme selfishness may come to deprive the individual of social services which at any moment may become essential to comfort, health, or even to life itself. It may be inferred that even among chimpanzees a certain degree of unselfishness is profitable.[9]

Several years later, Yerkes's associates at the Yale Primate Research Center, Henry W. Nissen and M.P. Crawford, added to the mounting evidence that chimpanzees engaged in "altruistic" acts. Nissen and Crawford defined altruism as an act that required the individual to sacrifice some resource for another's benefit. In their study, the resource was food.

Food sharing among chimpanzees was a sophisticated form of altruism, they found. More primitive altruistic acts involved an immediate exchange of goods or favors, as in barter, teamwork, or prostitution. Sexual desire played an interesting role in this form of primate reciprocity. The authors noted that: "Among some of the lesser primates especially, prostitution, [i.e.] exchange of food or freedom from attack for sexual favors, seems to be not uncommon."[10]

Delayed exchange was characteristic of a more complex form of altruism. Here, individuals had to keep a mental record of previous benefits received from specific individuals in order to decide with whom they would share resources. The authors described this delayed sharing as essentially, "barter on a 'credit' or deferred basis."[11]

Food sharing among chimpanzee "friends" was an excellent example of the existence of this higher form of altruism among nonhuman primates. They noted that a chimpanzee might beg often, and successfully, from certain partners, though practically never from others. "In general, our subjects responded better to the begging of a friend than to that of an animal with whom they were not otherwise intimate."[12]

Nissen and Crawford also observed that "when chimpanzees give a certain call, which evidently indicates fear or distress, at least one of its companions almost invariably will rush to its side, put an arm around it, and in general comfort it and give it sympathy."[13] Crawford had also seen chimpanzees join and coordinate forces to reach a common objective.[14]

In line with the prevailing tone set by Yerkes at the Primate Research Center, Nissen and Crawford attributed chimpanzee sharing to the actualization of a hereditary instinct. "There are no indications that the pattern of begging behavior is individually acquired; the evidence points to its determination by genetic factors," the authors claimed. However, what led them to this conclusion is not specified.[15]

Altruism and Kinship Selection

Of all the areas of study later investigated by evolutionary psychologists, the theory of kinship selection and consequent altruism was least anticipated by the earlier sociobiologists and evolutionary psychologists. Earlier development of Hamilton's thesis seemed to elude the original sociobiologists and evolutionary psychologists because of their continued reliance on Darwin's idea of group selection for an explanation of altruism.

Nevertheless, there were a few instances in which several early evolutionary psychologists seemed vaguely aware that altruism might be involved in increasing the survivability of kin. Edward A. Ross, one of the foremost American sociologists of the interwar period,[16] saw the foundations of altruistic behavior in the care that parents gave their offspring. Parental expressions of "tenderness" and "sympathy" for their young increased the likelihood that the parents' genes would survive into future generations, and so predominate in the population over time.[17]

Charles Ellwood, a highly respected sociologist at the University of Missouri, agreed that altruism sprang from the "parental instinct." He further speculated that an animal which could sense "points of similarity" between itself and other individuals of the same species might be instinctually driven to engage in altruistic behavior beneficial to the latter.[18]

Due to the affects of parental altruism, Samuel Holmes found it instructive to conceive of evolutionary struggle as guiding the destinies of entire families, rather than individuals:

> Nature, in endowing animals with instincts for maintaining life, is not interested particularly in the individual. She is concerned, if I may speak figuratively, with what Mr. Galton has called the stirp, or line of genetic

connection. In the struggle for existence the stirps mingle and compete, and natural selection eliminates some and preserves others. Natural selection effects evolution not through the removal of individuals as such, but through eliminating or preserving stirps. The stirp is the only concern of natural selection. It is the chief concern of accessory reproductive activities, such as parental care. It is the chief concern of infra-human morality. It is the concern, also, of most primitive human morality . . . [19]

Curiously, Roswell Johnson suggested that humans might employ a sort of "kinship selection," among other criteria, when contemplating marriage. He believed that the "accomplishment[s] and longevity" of a prospective mate's kin influenced mate selection.[20] Though not strictly an example of altruism, this idea did hint at an awareness that selection and preferential treatment might not be guided solely by the merits of one individual, but could include consideration of genetically related kin as well.

Of all the pre–World War II evolutionary biologists, J.B.S. Haldane came closest to discovering the theory of inclusive fitness. As he put it in his classic work, *The Causes of Evolution*: "For in so far as it makes for the survival of one's descendants and near relations, altruistic behaviour is a kind of Darwinian fitness, and may be expected to spread as a result of natural selection."[21] Haldane did not develop his insight further, however. An explication of the mathematical formulae relating the degree of shared genes between the benefactor and the recipient of an altruistic act, and the likelihood that self-sacrifice would benefit the survival of the shared genes more than would selfish behavior, would have to await the much later work of William Hamilton.

Rape

Altruism, then, could enhance the survivability of an individual's genes. So could rape, unfortunately. Mere days before World War II was unleashed, Samuel Holmes put the matter in stark Social Darwinist terms, in his address to the Western Division of the American Academy of Science. In nature, Holmes pointed out, "Cruelty, selfishness, lust, cowardice and deceit are normal ingredients of human nature which have their useful role in the struggle for existence. Intrinsically they are all virtues." For Darwin, Holmes explained, what was moral was what allowed the organism to survive and reproduce healthy offspring.[22] The brutal amorality of this vantage point could rationalize the act of rape.

Geritt Miller, Jr., was perhaps the evolutionary psychologist most concerned with the biological foundations of rape. Naturally, he approached the topic from the viewpoint of comparative psychology. He noted that acts of rape had been observed in chimpanzees. Males would occasionally use their greater physical force to overpower a female and force her into a sexual act. The male sex drive, which impelled him to seek sex with as many attractive females as possible, provided the motive for this behavior. Miller was well aware that this reprehensible behavior applied to humans as well as to chimpanzees. He somberly remarked that "women suffered from the accepted fiction that they are free to choose [their sexual partners] and the never entirely forgotten reality that in the last appeal they are not." Miller also found evidence for the existence of sexual masochism in primates, based on the work of Yerkes's colleague Harold C. Bingham.[23]

Holmes and Roswell Johnson wrote that sexual slavery, like rape, was another unfortunate consequence of the male sex drive. In past epochs, successful warriors were often awarded women from a defeated tribe, or even from his own, "with or without her consent." Such men, if they had many wives, were apt to have many children, and so spread their genetic predilection for aggression throughout the population.[24]

Art

Though it could be argued that male sexual instinct might lead in some cases to rape, no one believed that this was most often the case. As we have said before, most evolutionary psychologists agreed that women were usually free to choose their mates. Though women might admire aggressive men, Charles Ellwood, Roswell Johnson, and others thought that women would most likely choose men who attracted their attention through aesthetic display.[25] A man could use his intellect to enhance his physique with art, or produce artistic works that displayed his intelligence to prospective mates. Intelligence, it was assumed, might herald the ability to acquire resources and to defeat enemies, both traits that a woman would value. In several of his pre–Great War writings, Roswell Johnson speculated that "many of our esthetic attributes, such as musical and artistic ability . . . have been produced by sexual selection."[26]

In 1914, Samuel Holmes essentially anticipated the much later hypothesis of evolutionary psychologists that sexual selection might even account for the highly evolved human mind. His *Popular Science Monthly* article with the suggestive title, "The Role of Sex in the Evolution of Mind" emphasized, for example, that the remarkable ability of humans to communicate through the intricate sounds of speech may have evolved as a secondary consequence of vocal and hearing organs originally used to locate and identify potential mates. Holmes summed up the importance of sex for mental development as follows:

> The necessity for mating has, in general, been a constant force making for the evolution of activity, enterprise, acuity of sense, prowess in battle, and the higher psychic powers. We cannot pretend accurately to gauge its role in the evolution of the mind, but it has evidently been a factor of enormous potency.[27]

McDougall was also interested in the relationship of sex to mental evolution. He realized that young males could use aesthetic activities to attract potential mates:

> Dance and song and the writing of love letters, which figure so largely in the arts of courtship, connect the large fields of social activity in which the influence of the sex impulse is very obvious, with an equally extensive and perhaps even more important province of human activity in which the influence of the sex instinct is more obscure but undoubtedly present, namely, the production and enjoyment of works of art. The dance and song and literary composition which are used more or less deliberately in courtship may clearly be brought under the general principle that the cognitive energy of the instinct maintains all activities that appear to be means towards the attainment of the instinctive end. In this respect they are comparable to the efforts of the young man to secure an economic position which will enable him to marry the girl of his choice; efforts which, as we know, are often very energetic and long sustained.[28]

In the 1990s, Geoffrey Miller would rediscover the theory that sexual selection could account for the evolution of human artistic and intellectual creativity.[29]

Religion

As human intelligence evolved, humans became capable of communicating through symbols. This allowed abstract ideas to develop. Humans became anxious to understand the forces of nature, and wished to divine their relationship with these forces. They also sought to understand and control their own instincts. After all, many human instincts exist in precarious tension with the demands of civilized society. Several evolutionary psychologists considered this as the origin of religion. The imperatives of divinely ordained "morality" might explain kinship altruism, or could be invoked to curb male's polygamous desires when they threatened the social order.[30] Samuel Holmes made similar points in his unpublished book, *The Ethics of Enmity*.[31]

Of course, some scientists would have considered such speculation to be much too far in advance of the actual state of knowledge on the biology of human behavior that existed by the 1930s. There was the real risk that critics of evolutionary psychology would find these conjectures transgressed the proper bounds of biological inquiry.

*   *   *

Had a contemporary observer surveyed the field of evolutionary psychology in the early 1930s, he or she might have every reason to conclude that the foundations of a new scientific understanding of human behavior had been laid. The very significant work of a small coterie of biologists and social scientists, who were devoted to investigating the implications of Darwin's theory of sexual selection, had uncovered the motivations driving human sexual behavior. They had also offered provocative explanations for some of the most important elements of the world's major moral systems, which they now thought of as the consequences of evolutionary imperatives.

This by no means suggests that evolutionary psychologists had exhausted the potential of their field. The possible influence of heredity on a vast range of human behaviors had hardly been considered. Even where some hypothetical links between heredity and behavior had been postulated, they more often than not still lacked validation through empirical studies. Still, in an ideal world, the work done to date should have served as the basis for much more detailed and advanced studies of human evolutionary behavior in the late 1930s and thereafter.

However, many other forces were at work here than the mere quest for knowledge on both sides of the nature–nurture divide. The more tenuous hypotheses of the evolutionary psychologists concerning the influence of evolution on the development of human ethical behavior would prove to be particularly vulnerable to attacks by those whose ideological proclivities made them especially hostile to the suggestion that some human behaviors were influenced by evolution and heredity.

Part III

The Death of Sociobiology and Evolutionary Psychology

6

The Rise of Environmental Behaviorism

No science should go beyond the descriptive level. Specific stimuli determine specific responses; given a stimulus, a definite response can be predicted. What else do we need besides this for the scientific description of behavior?

Zing Yang Kuo, "The Net Result of the Anti-Heredity Movement in Psychology," p. 191

Nature only answers those questions which we ask her; indeed, she only gives the observer those answers which he expects from her.

Friedrich Alverdes[1]

[I]f our conception of human nature is to be altered, it must be by means of truths conforming to the canons of scientific evidence and not a new dogma however devoutly wished for.

Edward O. Wilson, On Human Nature, *p. 35*

Introduction

Science is influenced by the society that produces it. It may be that it is impossible for human beings to directly perceive reality without some degree of mental filtering and processing that shapes their perception. Certainly, observers differ in their interpretation of nature. This may be due, in part, to temperament, to background, to conviction, or to training. Institutional culture and historical trends also influence the questions scientists ask of nature, and the interpretations that they create. All these factors played significant roles in the struggle between evolutionary psychologists and environmental behaviorists to dominate science. In this case, by 1930 historical events had sufficient impact on the debate to tip the balance in favor of the environmental behaviorists. Thereafter, the affects of the Depression and the reaction against Nazi racial hygiene would fuel a thoroughgoing repudiation of evolutionary psychology. This chapter will attempt to explain how individuals, institutions, and historical forces interacted to lead to the death of the nascent sciences of sociobiology and evolutionary psychology, and to the triumph of environmental behaviorism as the sole acceptable interpretation of human nature for several generations.

In general, the institutional culture of American science, even before the pivotal decade of the 1920s, made hereditarian interpretations of human behavior increasingly difficult to maintain. There were two prominent reasons for this: the social sciences were demanding the adoption of more rigorous experimental methodologies, modeled after the "hard" sciences of biology, chemistry, and physics. At the same time (and with a certain irony) the social sciences sought to separate themselves from biology and philosophy, and sought an acknowledgment of their worth as independent disciplines from scientific and governmental institutions. The early twentieth century was a period in which such fields as psychology, anthropology, and sociology were organizing themselves and creating the typical "academic infrastructure" of university departments, textbooks, journals, national associations, and funding agencies.[2] To raise both interest levels and funds for these endeavors, the social sciences had to demonstrate their "worth" as distinct fields of study.

The eugenicists would try to deny them the opportunity to do so, thus helping to perpetuate the antagonism between environmental behaviorists and evolutionary psychologists. In one example, the eugenics leader Clarence G. Campbell wrote to Charles Davenport in 1928: "It appears to me that eugenics is the particular biological lever of which all the social sciences are in need in order to orient *their* policy."[3] Campbell expressed a similar domineering attitude to Lorande Loss Woodruff, the chairman of the Division of Biology and Agriculture of the National Research Council: "In fine, Eugenics seeks to discover all the means of improving the racial condition from a biological basis, a basis upon which all sociological theory and action must eventually rest."[4]

Social scientists had to contend with the tension inherent in proving that their sciences were founded on "provable" concepts different from those of biology; yet at the same time they were as "scientific" as biology.[5] This would not be possible if they relied on hereditarian explanations for human psychological and social behavior. Heredity was, after all, the most basic premise of biological evolution. Thus, the pioneers in the American social sciences discovered a new reality: that human behavior was actually based on learned patterns of activity passed down through the generations by language; in short, "culture."[6]

The Rise of Environmental Behaviorism

Franz Boas was the father of American anthropology as well as of modern environmental behaviorism.[7] Boas was from a German Jewish family of well-educated professionals. Early on, he found himself attracted to both physics and geography, but travels to Baffinland brought him in close contact with the Eskimo. He quickly became so fascinated with studying the Eskimo that he forgot physics, and immersed himself in what we would today call cultural anthropology. Anthropology was then, in the 1880s, just beginning to emerge as a distinct discipline.

Boas was wedded to his German identity, but his liberal Jewish values conflicted with conservative Wilhelmine Germany. Fortunately, Boas decided to seek his academic fortune in the United States, a more open society where he hoped his academic career would more likely flourish.[8] Several years after his expeditions to Alaska and to British Columbia, Boas succeeded in landing an academic position at Clark University, under G. Stanley Hall. It is likely that Boas had already developed a mindset that predisposed him to see human behavior as a product of culture, rather than of biology; certainly that is what he concluded from his early studies.[9] Wisely, he attacked hereditarian concepts of human behavior at their most dubious point, on racial differences. He saw no discontinuities between the fundamental behavior and intellect of peoples with less developed cultures and those of more complex cultures. He reported these findings in *The Mind of Primitive Man*, published in 1911. Boas supplemented this seminal text in cultural anthropology by one equally important to physical anthropology, *Changes in the Bodily Form of Descendants of Immigrants*. This work, part of a large government financed study, asserted that the cranial shape of European immigrants was not inherited by their descendants, but changed after exposure to the American environment.[10] Although perhaps arcane to the layman, the work refuted the idea, almost universally accepted by anthropologists, that cranial shape was one of the principle determinants of race. In essence, Boas's studies showed that genetic inheritance was not particularly important to understanding human behavior; race had no importance whatsoever.[11] The study's analysis of the data was met with criticism by eugenicists,[12] but they were unable to change the impression Boas had created that the common assumptions on man's genetic inheritance might be illusory. Boas was determined to spread these ideas throughout the growing anthropological community, not only by his own efforts, but especially by the efforts of his students.[13]

Coalition building was undoubtedly one of Boas's most notable skills. By educating a large number of graduate students to accept and promote his vision of anthropology, and by gaining control of important journals and professional associations, Boas and his students would dominate the field of anthropology by the 1920s. Since, at roughly the same time, followers of John B. Watson were gaining predominant influence in psychology, the social sciences became increasingly unified around the basic concepts of environmental behaviorism by the early 1930s.

Boas was fortunate in that the trend toward the increasing academic professionalization of anthropology favored university-trained anthropologists. Up to the turn of the twentieth century, anthropology was essentially the preserve of academic "crossovers" from philosophy (such as William James) or amateur anthropologists without university training in that subject, such as Daniel G. Brinton or John Wesley Powell.[14] In an effort to enhance the standing of the field, however, academic anthropologists insisted on "squeezing out" these nonprofessionals in favor of those with Ph.D.s in anthropology.

The first decade of the twentieth century was good to Boas: he advanced to a professorship at Columbia University, and was even elected to the presidency of the American Anthropological Association in 1907. As his students gained positions on the Association's governing council, the Boas faction gained increasing influence over that organization by the end of the decade. By 1915, the "Boasians" also controlled the Association's journal, *American Anthropologist.* Articles supporting biological determinism subsequently disappeared from the journal.[15]

Boas's students at the time tended to be the best and brightest of the new generation of anthropologists. These included the leaders of American anthropology in the succeeding decades: Alfred Kroeber, Robert Lowie, Edward Sapir, Alexander Goldenweiser, Paul Radin, Leslie Spier, Ruth Benedict, Melville Herskovits, and Ashley Montagu. They were all hostile, to varying degrees, to the assertion that human behavior was a product of biological heredity. Human behavior was the product of culture, pure and simple.[16] Boas's students soon became leaders of the major anthropology departments of the United States, and so set the research agenda and teaching paradigm for the discipline.[17]

To demonstrate the validity of environmental behaviorism, Boas's students scoured the globe looking for cultures that differed widely on even seemingly basic social behaviors. Most famous among them were Ruth Benedict, whose studies of Native Americans included *Concept of the Guardian Spirit in North America* (1923), *Patterns of Culture* (1934), and *Zuni Mythology* (1935). Benedict showed that culture rather than biology played the leading role in shaping societies. In primitive man, she explained, "not one item of his tribal social organization, of his language, of his local religion, is carried in his germ cells."[18] Benedict's student, Margaret Mead, followed in the Boasian tradition. Mead wrote *Coming of*

Age in Samoa (1928), *Growing Up in New Guinea* (1930), *The Changing Culture of an Indian Tribe* (1932), and *Sex and Temperament in Three Primitive Societies* (1935) before World War II. Mead appeared to discover Pacific island societies in which women were at least as powerful as men, and sexual conflict was rare. Thus the general public learned that nestled away in exotic tropical lands were societies free of the problems that seemed to plague Western cultures: sexual competition, jealousy, warfare, and selfishness. Given the obvious dichotomies with Western society, Mead and her associates suggested that there were no essential commonalities in human societies, and hence no innate behaviors. The ills of Western society might not be the result of innate behavior, more and more people dared hope, but of a society in dire need of reform.[19]

Unfortunately, it was learned in later years that both Benedict and Mead had presented disturbingly selective and limited data, thus skewing their conclusions. Later anthropologists would find that, for instance, the Samoans studied by Mead actually lived in a much more commonplace society marked by the same violence and patriarchical sexism as were most others.[20]

The extent to which the Boasians denied that humans were part of the natural, evolutionary world is in some instances breathtaking. Kroeber, for example, claimed in 1917 that "The distinction between animal and man which counts is not that of the physical or the mental, which is one of relative degree, but that of the organic [of animals] and the social [of humans], which is one of kind."[21] People had to realize, Kroeber asserted, that "while men are animals, animals are not men, and that however much a human being may have of the nature of the pig, he nevertheless has one thing that no pig ever had, namely the faculty for civilization and hence for morality . . ."[22] Indeed, Robert Boakes has even argued that the Boasians' zealous mission to spread the gospel of man's uniqueness and freedom from innate predilections was almost religious in nature.[23]

Others joined Kroeber in remarkably similar proclamations. For example, Lester Ward's *Outlines of Sociology* denied the animal basis of human society. As Ward put it: "It [human society] is essentially rational and artificial, while animal association is essentially instinctive and natural."[24] George Murdock in 1932 announced what had by now become social science dogma: "culture . . . [is] a uniquely human phenomenon independent of the laws of biology and psychology . . ."[25]

The Boasians were especially hostile toward the eugenicists, who they often condemned (with justification in many instances) as Anglo-Saxon elitist snobs, a few of whom did not even hold a Ph.D. in the social sciences or had accomplished any scientific research.

Initially, the eugenicists fought Boas with a passion equal to his own. Given that before World War I they were merely advocating beliefs already widely held in the academic community—that much of human behavior was based on heredity and that eugenics was a valid means to improve the human species—they garnered substantial support among their fellow academics. Though Boas's troops may have been gaining control over the American Anthropological Association, the hereditarians gained control over that most critical resource of modern science: money. The National Research Council, set up by prominent scientists during World War I to aid the war effort, established an Anthropological Committee in 1917 that was dominated by eugenicists, including Madison Grant, Charles Davenport, Ales Hrdlicka, and Robert Yerkes.[26]

The mutual dislike between Boas and Grant must truly rank as one of the great academic hatreds of the twentieth century. These men were polar opposites: Boas, a liberal German-Jew convinced of the predominance of environment in determining the human character; and Grant, a wealthy Anglo-Saxon Protestant who disliked foreigners (Jews in particular), admired Adolf Hitler, and devoted his life to advancing the causes of immigration restriction and radical eugenics.

Grant would tell anyone willing to listen how much he loathed Boas. He explained to his editor at Scribner's Publishing House, Maxwell Perkins, that Boas "naturally does not take stock in any anthropology which relegates him and his race to the inferior position that they have occupied throughout recorded history."[27] At a meeting of the Galton Society, Grant waxed eloquent in his sermon on the true object of science: not the "study of pottery and blankets" indulged in by Boas and his followers, but "the study of man as a *physical* animal," "the naked *anthropos*."[28] Boas, for his part, had nothing but contempt for Grant's "Nordic nonsense," not to mention his lack of appropriate scientific credentials.[29]

World War I was something of a God-send for those who opposed Boas. They turned the anti-German, anti-immigrant hysteria of the period against Boas in particular and the immigrants he was sympathetic toward in general. Boas, for his part, was not one to keep his opinions to himself. He was irritated by the war, concerned about its negative impact on Germany, and infuriated by the restrictions on civil rights that the war engendered. When Columbia University president Nicholas Butler suspended free speech at the university for the duration of the war, Boas retaliated by writing an open letter denouncing the high-handed treatment of faculty rights.

Boas's enemies were delighted. They maneuvered the American Anthropology Association into censoring him, in 1919, for his "anti-American" statements. However, the ideological fervor of wartime died down as the true murkiness of the war became clear. By the mid-1920s, Boas was free from any restrictions, and indeed more admired than ever.[30]

As the Boasians were gaining control over American anthropology, a similar paradigmic shift was taking place in American psychology. Most members of the discipline came to conclusions similar to those of anthropologists: that complex behavior was learned. The impetus for this transformation came from a rather quirky psychologist at Johns Hopkins University, John B. Watson. Watson was originally a close friend of Robert Yerkes, and like him was fascinated by animal behavior. These men took different tracts, however, in their approaches to their subject on the eve of the World War. Yerkes continued to probe into the inner workings of the animal mind. He was driven to understand to what extent humans and chimpanzees shared cognitive abilities and processes, to what extent chimpanzees "think" and are conscious.

Watson, like some of his young colleagues in psychology, was coming around to the conclusion that these questions were senseless philosophical speculations. It was simply not possible to get into an animal's "mind"; indeed, later behaviorists such as B.F. Skinner would conclude that it is impossible to get into *any mind*, including that of a human being. Watson became convinced that psychology, as a science, should strive to accumulate empirical evidence concerning the "stimulus–response" relationship of behavior (referred to contemptuously by William McDougall as "muscle and twitch" psychology). The experimenter provides a stimulus to affect the behavior of a child or a chimpanzee, for example, and then measures the behavioral response. This was true, objective science, shorn of metaphysical fantasy.[31] Watson's masterpiece, *Behaviorism*, published in 1924, would grow almost poetic in its assertion that there was no behavioral difference of any significant sort in newborns. Virtually all behavior was learned, Watson announced. Such concepts as innate talents or abilities, indeed the very concept of innate predispositions to certain traits, were abolished. Watson had no use, either, of such quaint notions as mind, ideas, beliefs, desires, or feelings. How could such vague concepts ever be measured? Since science was based on variables that could be measured, these concepts had no scientific existence, anymore than did God or soul.[32] Watson offered such circumlocutions to explain behavior as: "Can we not say that he [man] is built of certain material put together in certain complex ways, and as a corollary of the way he is put together and of the material out of which he is made—he must act (until learning has reshaped him) as he does act?"[33]

Watson quickly developed a following in the years immediately following World War I. There are several reasons for this. For one, Watson's behaviorism was born and bred in the United States. It did not suffer from any association with "foreign" (and especially German) psychological theories, which was a consideration in the isolationist atmosphere of the

1920s. Behaviorism also corresponded well with logical positivism, which was a popular philosophical system of the time. Logical positivism demanded that theoretical concepts be closely associated with the operations used to measure them. Watson would have heartily agreed. Likewise, Watson's behaviorism narrowly focused theory down to observable measurements of behavior.[34]

Boas's followers were adamantly opposed to the concept of human culture having biological foundations; Watson's followers were positively contemptuous of anyone continuing to suggest that human behavior had instinctual roots. Zing Yang Kuo, an American-educated Chinese psychologist, was one of the more extreme pioneers of the new doctrine of behaviorism. In fact, a horrified McDougall described him as "out Watsoning Watson."[35] Kuo was convinced that instinct was a "finished" idea in psychology.[36] So fierce was Kuo in his convictions that he was even moved, in 1929, to intemperately declare: "All our sexual appetites are the result of social stimulation. The organism possesses no ready-made reaction to the other sex, any more than it possesses innate ideas." Behavior was determined by only two factors: the neurological structure of the organism, and its environmental stimuli.[37]

Although B.F. Skinner may have become more famous than Kuo, he was a no less determined opponent of evolutionary psychology. Indeed, Skinner was the father of "radical behaviorism." He developed the concept of operant conditioning: all behavior can be explained through the conditioning of the organism in response to the receipt of rewards or punishments for its actions. Reward it, and it will be more likely to repeat the behavior. Punish it, and the behavior will be reduced. Human beings had one and only one innate behavior: they could learn. As it would soon be summed up, "The nature of human nature is that humans have no nature."[38] Human beings, and by implication human society, could mold themselves to become whatever they willed.

By the mid-1920s, the twin currents of culture-oriented anthropology and behavioristically oriented psychology joined to form the torrent of environmental behaviorism that would soon sweep over the American academic world. Both groups were united in supplanting the concept of instinct with that of "habit"; and in stressing the greater explanatory power of cultural and social environment over biology in understanding human behavior. Both had sneaking suspicions that biological explanations of human behavior were inherently racist.[39]

The dwindling number of eugenicists and evolutionary psychologists were aghast at the spread of environmental behaviorism. Not only did it defy scientific evidence, it defied common sense. The evolutionists fought Boas, Watson, and their followers with increasing frustration, and decreasing success, until World War II. David L. Krantz and David Allen found that there were at least fifty attacks against each side in the scientific journals from the 1919 to 1930.[40]

Julian Huxley would initially be one of the most outspoken critics of behaviorism, which is not surprising, since his reputation was built on work describing the evolution of instinct and emotions. In a 1923 article, Huxley condemned behaviorism's dismissal of subjective psychology as "unscientific." The "principle of uniformity" supported the view that emotions existed in both animals and humans. Given the great similarity of behavior in some cases, to claim a whole class of behavioral phenomena existed in animals, but not in humans, was "to make scientific reasoning a farce."[41] In the same essay collection, Huxley affirmed his allegiance to hardcore eugenics. He looked forward to the day when eugenics would be enshrined in "practical politics." Then it would be applied to raising the average genetic quality of the population, "by altering the proportion of good and bad stock, and if possible eliminating the lowest strata, in a genetically mixed population."[42]

Naturally, the combative William McDougall was deeply offended by Watson and his associates. McDougall was the chief defender of instinct theory in the United States after accepting an appointment at Harvard in 1920, and often debated Watson on behaviorism. McDougall argued that the obvious existence of instincts in animals left no doubt that they must also exist in humans. He made his attitude regarding the new scientific approach quite plain: ". . . the present endeavor of some biologists and psychologists to abolish the mystery of heredity by denying it all but a very slight influence on human and animal behavior seems to me ridiculous."[43] In a private letter to Duke University president William Few, McDougall described behaviorism as a "foolish and pernicious perversion of philosophy . . ."[44]

McDougall helped poison his scientific reputation by so violently reacting against behaviorism that he forced himself into an almost equally extreme and indefensible position. He believed that any behavior was a sign of consciousness, no matter how rudimentary. This, McDougall admitted, applied across the animal kingdom, even to amoeba. McDougall also concluded that the animal mind could never be fully explicable without reference to a nonmaterial energy, a sort of "soul."[45] This belief led him to pursue the study of psychic powers, such as extrasensory perception. So fascinated was McDougall with the "mysteries" of the body–mind connection that he even dabbled in attempts to demonstrate the existence of Lamarckianism (the belief that physical changes occurring during an organism's lifetime will pass into its genes and be inherited by its offspring). These rather bizarre ideas, which McDougall never ceased to publicly endorse, discouraged his colleagues from taking him seriously. As a result, McDougall's much more sound insight into animal and human instinct was also tossed aside as just one more fanciful idea of Professor "Potty."[46]

McDougall's outspoken advocacy of eugenics and Nordic racial supremacy also worked against the acceptance of his evolutionary psychological ideas. In his 1921 book *Is America Safe for Democracy?* McDougall advocated the use of eugenics to preserve the New England strain of the Nordic race in America.[47] In a 1924 debate with Watson at the Annual Psychology Association meeting in Washington, D.C., McDougall claimed that his "hobby" of eugenics was one of the most important possible solutions to the great problems that confronted civilization.[48]

McDougall rapidly sank into obscurity after he left Harvard for Duke University in 1927.[49] McDougall became a bitter man. As he complained in a letter to the *Fortnightly Review*, with regard to yet another attack against him: "Since my exile to America I have continued almost alone amongst the psychologists here to fight against the various mechanical psychologies which still are prevalent . . ." This "fervid advocacy" of the "causal efficacy in nature of human intelligence and purpose" had only made McDougall "something of an outcast from the scientific world . . ."[50] To another correspondent McDougall mourned: "My outlook of the future is somewhat gloomy, and I strongly incline to feel that we are in a period of both moral and intellectual decline which is very fundamental . . ." This sorry state was due to "racial deterioration" brought on by the "disproportionate reproduction from the least well endowed part of the modern population . . ." Only McDougall's "insistence upon the need for far reaching eugenic measures" as he had expounded in his book *Religion and the Sciences of Life* offered any hope.[51] McDougall was quite aware that his views on eugenics, "problems which would seem to be regarded as strictly taboo by the greater part of the press and the men of science who ought to be interested in them," had also cost his reputation dearly.[52]

Besides McDougall, others dared attack Watson and other environmental behaviorists on the grounds that intelligence was inherited, not essentially a product of upbringing. Samuel Holmes maintained that "There is a vast amount of evidence that mental differences are inherited and that they follow the same laws of inheritance as physical differences."[53] Holmes took exception to "egalitarians" such as Chester Ward whose reasoning contained the "serious flaw" of explaining intellectual achievement by way of good environment: that smart kids were smart because they were raised in homes with plentiful educational and financial resources. Rather, the parents of such children most likely used their own innate intelligence to acquire such desirable resources, and so provide a stimulating environment for their children, who benefited from both their own innate intelligence and the good environment their parents could provide them. Holmes concluded: "Equalizing opportunity, as [Edward] Thorndike has shown, does not tend to equalize achievements . . . Improving opportunity increases instead of decreases the initial disparity between individuals. You can level human beings down. You cannot level them up."[54]

Several years later, Holmes was moved to analyze the errors of his "environmentalist" opponents. For one, they refused to accept the "substantial contribution to our knowledge of inheritance" made by the founder of eugenics, Francis Galton. In his magnum opus, *Hereditary Genius*, Galton used "impartial statistical methods" to "conclusively" prove that "superior mental ability runs in families, and that the more eminent a person is the greater the number of eminent relatives who will on the average be found in his family." Ignoring Galton's work, environmentalists continued to challenge the "hereditarians" on the origins of intelligence and other traits. Whenever anyone claimed that a given difference in mentality was due to heredity "the environmentalist may challenge . . . [the] statement and claim that it may be due to environment, or at least that we can not prove that it is not."

John B. Watson was the "most extreme environmentalist." Watson believed that "'there is no such thing as an inheritance of capacity, talent, temperament, mental constitution and characteristics. These things again depend on training that goes on mainly in the cradle.'" We only inherit structures, Watson claimed. "'Habit formation starts in all probability in embryonic life and that even in the human young, environment shapes behavior so quickly that all the older ideas about what types of behavior are inherited and what are learned break down.'" Holmes sarcastically concludes from this that Watson's "conception of heredity seems to be very different from that of the modern geneticist." Watson "is led to adopt a position which is open to all the objections which have been urged against the discarded teachings of Locke. For him the mind is a *tabula rasa* upon which experience writes all the contents."

To Holmes, Watson and his followers were so deluded that they seemed to think nothing of making preposterous claims. "Some even go to the ridiculous length of questioning whether the laws of heredity apply to man at all!" the outraged Holmes remarked. "This is merely wishful thinking on the part of those who prefer to bask in a fool's paradise. They will not face the overwhelming evidence, but prefer to ignore facts and pretend that every human mind has unlimited possibilities and that heredity plays no part in its development."[55]

How had such thinking infected science, Holmes wondered? He decided that "a considerable number of those concerned with psychology, education and the social sciences have a very imperfect grounding in the principles of modern genetics." Furthermore, they were loaded down with biases:

> we have various kinds of emotional bias which influence opinions on the subject to a marked degree. There is the theological bias which leads people to regard with disfavor the doctrine that traits of the mind obey the same laws of heredity that obtain for the material body. A probably stronger bias in these days arises from the varieties of political and social philosophy whose votaries scent a danger in the doctrine of the natural inequality of man. We have also what may be called the humanitarian bias, commonly found among those engaged in the uplift of their fellow creatures, which predisposes people to attribute human ills, so far as possible, to remediable causes. The so-called fatalistic teaching of the hereditarians is regarded as a sort of challenge to the efficacy of their efforts at social improvement . . . As a result of many kinds of bias, complexes, phobias and idols we have a large body of interested and often aggressively good people who are eager to welcome anything which seems to weaken the position of the hereditarians.

These biases blinded the environmentalists to the reality that many human personality traits were inherited. Holmes can only conclude that "The environmentalist argument is very much like that of a gardener who would claim that since soil and cultivation make such striking differences in the growth of plants, the matter of possible differences in seed can be safely ignored."[56]

Some of the comparative psychologists also refused to accept Watson's thesis. Watson and Yerkes, who had been colleagues and friends, parted company in 1913. Yerkes was dismayed at Watson's announcement of behaviorism in his 1913 paper to the American Psychological Association. Watson, for his part, thought Yerkes did sloppy work.[57]

In 1922, however, Yerkes was taken aback at Walter Lippman's series of articles in the *New Republic* attacking the validity of the Army IQ tests that Yerkes had directed.[58] Always desirous of avoiding open controversy, Yerkes abandoned further research on inheritable traits in humans. His public commitments to evolutionary psychology and eugenics, though never wavering, were somewhat more cautious after the early 1920s. He made perhaps his strongest statement on the broader dispute in *Almost Human*: "the person, whether layman or scientist, who depends wholly on rigidly controlled laboratory studies for his knowledge of the anthropoid apes or of man is naïve indeed, and to be pitied rather than abused." He considered the experimental methods of the behaviorists "often over-refined and artificial."[59]

Yerkes continued to express his more controversial thoughts in private correspondence, however. In 1926, Yerkes congratulated Henry Pratt

Fairchild on his new anti-immigrant book, *The Melting Pot Mistake.* Yerkes told him, "In general, I think we are in agreement . . . I like the book very much, am heartily glad that you have written it, and hope that it may be read thoughtfully and sympathetically by millions of our citizens."[60]

Participating in an American Eugenics Society campaign to require noncitizen immigrants to register with the U.S. government, Yerkes wrote to President Herbert Hoover:

> Presuming on my experience in dealing with certain scientific aspects of our problems in immigration, I express the hope that you may be able to give serious attention to the matter of immigration and may be led to conclude, as have I, that wise restrictive measures are of the utmost practical importance for aliens as well as for our citizenry. . . . I am heartily endorsing the enclosure as prepared by Committee on Selective Immigration of the American Eugenics Society. . . . Particularly I would urge on your consideration the proposal for the registration of alien population. It is my personal conviction that we should have universal registration, and that under existing conditions the least we can do, as an intelligent citizenry, is to provide for the registration of all aliens.[61]

Yerkes did wish to make it clear to his colleagues, however, that he was committed to an understanding of psychology firmly based on biology. Thus, in the late 1920s he formally adopted the title of "psychobiologist."[62] He told Edwin G. Boring that the term meant "the study of consciousness as a biological phenomenon and of its associated organic activities."[63]

Yerkes also continued to see his primate research intimately connected with understanding human psychology and having eugenic overtones. These relationships were spelled out by John C. Merriam in his 1930 address to the American Anthropological Association, "The Practical Significance of Studies in Early Human History":

> There is a word that is used commonly these days that is spelled "eugenics." It is in a general way the idea that we have the right to be well-born. I conceive of it also as meaning that people have the right to be better and better born through the ages. Not less important than that which has come to all generation preceding us, the subject of eugenics must be considered from many points of view. We know something of the new Institute for Human Relations which is now being established at Yale University, a very great thing, the interrelation of studies that have to do with anthropoids, with the intimate structure and physiology of man, the relation of it all to laws of human conduct. It will not be possible to solve all the problems at Yale, nor do they expect to. They will not solve them all with that one institute, or in this century, but we are making a beginning, one of many beginnings of a

study that has as its object this greater thing—the understanding of our capacity and the means of improvement of mankind in future generations.[64]

Yerkes's other eugenicist friends also deplored the victory of environmental behaviorism. Lewis Terman, who worked with Yerkes on the Army IQ tests, called Watson's behaviorism "a cult." In his autobiography, Terman wrote that behaviorism's "presumption in claiming the whole of psychology and in basing a theory of child training and a denial of heredity on a few minor experiments in the emotional conditioning of infants is ridiculous."[65]

Franz Boas was just as culpable, according to the evolutionary psychologists. Clark Wissler, who served with Yerkes and Merriam on the Carnegie Institute's Advisory Committee on Research on Human Behavior in the early 1920s, would later tell a colleague that he had "never sympathized with Boas' deep hostility to biological science," and was well aware that Boas considered him "a heretic."[66]

In the late 1920s, Gerrit Miller Jr. boldly attacked two prominent social scientists, J. Arthur Thompson and Bronislaw Malinowski, both inclined toward environmentalist explanations of human society. Miller wrote with obvious irritation that, because of their writings, "it should not surprise any one ... to find how strongly, even in the technical literature of psychology and anthropology, the idea of man's essential peculiarity is entrenched."[67] Instinct in the lower primates, Thompson and Malinowski argued, disappeared in favor of culturally conditioned behavior in man. Miller believed very much the opposite: since there was little appreciable difference in sexual behavior between different primate species, including humans, then at least one of the most fundamental and important human activities, sexual behavior, was innate.[68]

In spite of the efforts of Yerkes, Miller, and others, evolutionary psychology was doomed. Its association with the odious politics of eugenics was frequently highlighted, and used to condemn both. It is important to note that new scientific discoveries were *not* essential reasons for this shift in perspective.[69] As Hamilton Cravens pointed out, the triumph of Watson's behaviorism over biologically oriented psychology offered neither a more logical explanation of human behavior, nor more experimental evidence.[70] Rather, the explanation for this startling reversal on the roots of human behavior must be sought in the complexity of American society and American science in the interwar years. There were social, methodological, institutional, and economic factors in this paradigmatic change.

7

Evolutionary Psychology under Attack

In the field of science, as in any other . . . individuals . . . find themselves behaving just as other human beings do in similar situations. In the scramble for places, for economic security and prestige, things are done and offenses committed against human beings whom . . . scientists are so selflessly supposed to serve, offenses which, if anything, are perhaps a little worse than those for which other classes of mankind are so frequently condemned.

M.F. Ashley-Montagu, "Selfish Scientists," p. E9

Eugenics became increasingly out of step with the changes occurring in American society in the interwar years. The desire to reform American society and provide opportunities for advancement to the less fortunate was temporarily neglected in the 1920s, but came back with full force with the challenges of the Depression era. Social scientists who wished to contribute to the restructuring of American society and increase its fairness and egalitarianism were put off by the elitist presumptions and racist overtones of eugenics. Rather, the environmental behaviorists offered a paradigm that seemed to suggest much more open possibilities in human potential. In essence, environmental behaviorism seemed to be more liberal, and a more fitting scientific ideology for liberal minded Americans.[1]

This perspective on the nature of American society was fully embraced by the new generation of social scientists becoming prominent in the 1930s. Many of these men and women were from immigrant backgrounds, and were much more ethnically diverse than were the American social scientists of the turn of the twentieth century. Their career success depended on the United States becoming a more fluid, tolerant, and culturally open society. Not surprisingly, they found environmental behaviorism much more amenable to these goals.[2]

It certainly escaped no one's attention that many advocates for evolutionary psychology were notable eugenicists, and several of the most outspoken eugenicists were blatant Nordic race supremacists. Eugenics offered a facile scientific garb to dress up the Nordicist racist agenda. The elitist implications of eugenics made it a seemingly modern science in search of a reactionary cause such as Nordicism.

As a result of these factors, by the early 1930s eugenics became more closely associated with conservative, even reactionary political and social views. Liberal eugenicists (and there were quite a few) either fled the movement or muted their views.[3] The rise of Nazism, and its poisonous embrace of eugenics, accelerated this development. The reaction against eugenics, and all associated with it, spread to include a condemnation of any assertion that genetics and human behavior could somehow be linked.

*   *   *

Continuing a trend that had begun in the turn of the twentieth century, the newer generation of scientists interested in human behavior wanted to see the results of rigorous experimentation before they bought into any explanation for human behavior. In their opinion, evolutionary psychology lacked any serious credibility. Being unaware of (or ignoring) the studies discussed in chapter 4, the environmental behaviorists saw only the assumption that animal and human behavior had related instinctual causes as grounds for accepting evolutionary psychology, and they regarded such a hypothesis as meritless.[4] If anything, they regarded eugenics as even more bereft of convincing experimental support.

Furthermore, as the science of genetics developed, it became apparent that the eugenic rush to mass manipulate human reproduction based on simple high-school genetics generalities was dangerously ill-founded. The simplistic "dominant-recessive" dichotomy of genetic expression beloved of eugenicists, which decorated their genealogical charts, could not explain numerous traits that were known to be inherited. How less likely was it that the eugenicists could account for subtler distinctions, such as intelligence and personality traits?

Knight Dunlap was one of the first evolutionary psychologists to cross over to the other side. In the first article of what would become many attacking instinct, Dunlap pointed out that evolutionary psychology's dependence on instinct theory committed the fallacy of nominal thinking— the tendency to confuse naming with explaining. This logical lapse, which was interpreted as covering up the fact that instinct theorists had no evidence for a physiological mechanism to explain their concept, helped fuel a revolt against instinct theory.[5] Since many eugenic theories depended on hereditary behavior often described as "instinctual," the reaction against instincts struck a blow against eugenics as well.

Those geneticists less transfixed by eugenics increasingly found it wise to distance themselves publicly from the pronouncements of its more extreme advocates. Between 1925 and 1930 Herbert Spencer Jennings, William Castle, Raymond Pearl, Thomas Hunt Morgan, and, finally, Edwin Conklin published articles critical of eugenics that took pains to explain to the public that eugenics was essentially political propaganda divorced from the true science of genetics.[6] Raymond Pearl, a sometime eugenicist who announced his break from the pack in a 1927 article published in *American Mercury*, got into a debate on the issue with his friend and loyal eugenicist, Harvard geneticist Edward Murray East. In a letter to East, Pearl explained his rationale for publicly speaking out against eugenics:

> You go on to say that naturally no modern geneticist believes in an extremely high correlation between characters and characteristics of parents and offspring. But that is exactly the point of my whole article—that no modern geneticist does believe in such a high correlation. That is precisely why I wrote the article; to emphasize the fact that no qualified geneticist does believe in it. It is only the brash eugenic boys who take that position, and they are the people I am after, not the geneticists. Could anything be plainer than this in the whole article? You seem to take the position that I am attacking geneticists. On the contrary, what I thought I was doing was defending genetics from its friends. You go on to say that I shall be severely criticized if I maintain this position. No doubt I shall be by the people whom I am attacking, but the paper about which you are talking was read by a number of the most distinguished geneticists in the world before it was printed, and approved of. In fact I was urged, and finally persuaded after refusing to do it, to present just the case I did at the Berlin Congress on behalf of geneticists.[7]

The famous eugenicist Charles Davenport's great project at the end of the 1920s—his *Race Crossing in Jamaica* study—laid bear the fallacies that penetrated eugenic research. Davenport was the chief researcher on the project, which was funded by William Draper, a wealthy contributor to various "pro-white" eugenic studies. Edward Thorndike, Clark Wissler, and W. V. Bingham served on the advisory committee. Morris Streggeda, a young researcher at the Eugenics Record Office, was sent to Jamaica for thirteen months to gather anthropometric and psychological data from people who were identified as white, black, and of mixed-race. Steggerda diligently gathered 5,000 pages of data on 78 different traits. While Steggerda was on his mission, Davenport grew concerned that the young researcher might miss the point of the whole project. Davenport reminded

Steggerda to keep focused on the study's "main problem, namely, the capacity of the Negro to carry on a white man's civilization." To this end, Steggerda was ordered to pay more attention to those psychological traits that really mattered, such as "rating[s] of honesty, thrift and foresight." These, after all, were the characteristics on which "a white man's civilization" was built, Davenport felt compelled to reiterate, as if Steggerda was a wayward student.[8]

Steggerda's final results did not give Davenport much solace. In sum, they did not show nearly the sorts of discrepancies between the intelligence of whites and blacks that Davenport had expected. Nor were people of mixed-race conflicted and incapable, as Davenport had for years predicted. Nevertheless, Davenport was grimly determined to explain the results in such a way as to support the conclusions he had arrived at long before. The unexpected finding that blacks performed better than whites at "complicated directions for doing things" as well as "in simple mental arithmetic and with numerical series" was confidently explained away by Davenport: ". . . the more complicated a brain, the more numerous its 'association fibers,' the less satisfactorily it performs the simple numerical problems which a calculating machine does so quickly and accurately." Davenport was initially surprised that "on the *average* the Browns did not do so badly" on the intelligence tests. But he was eventually inspired with the answer: more often than blacks or whites, mixed-race individuals "were muddled and wuzzle-headed." Why might this be the case? Because, Davenport continued, though blacks were of "low intelligence . . . they generally can use what they have in fairly effective fashion"; "Browns," on the other hand, "seem not to be able to utilize their native endowment." This was because the Browns suffered from biracial mental "disharmony."[9] Though Davenport's interpretation of the data may have satisfied himself, it permanently damaged the credibility of eugenics as a whole, given Davenport's status as chief spokesperson for eugenics in the United States. Eugenics seemed revealed for what it really was: an attempt to manipulate science into fitting predetermined conclusions that were actually inspired by an extremist political agenda. Davenport's *Race Crossing in Jamaica* was a milestone in the rejection of eugenics. Biologists such as William Castle reacted harshly against the obvious racial biases at the very foundation of Davenport's study. Davenport's own reputation as a biologist declined precipitously after the book's publication.[10]

Another example of Davenport's increasing irrelevance is evidenced by his stewardship of the Committee on Human Heredity of the National Research Council. On March 20, 1929, Davenport and Frederick Osborn, the nephew of Henry Fairfield Osborn, met with L.L. Woodruff, chairman of the Division of Biology and Agriculture of the Council, to petition for the inclusion of the Eugenics Research Association as a constituent member institution of the Division.[11] This proposal eventually was not accepted, but a compromise was reached. The National Research Council set up a Committee on Human Heredity, which had a twofold purpose: to fund projects that investigated the inheritable aspects of human psychology and physiology, and to act as a liaison between the Eugenics Research Association and the Council. Davenport was made chair, with Harry Laughlin, Clark Wissler, Edward Thorndike, and several others on the Committee.[12] With such a makeup, it is no surprise that Davenport's pet projects were advanced as the leading candidates for funding. He sought funding for studies of race crossing, and for genetic studies of "some special capacity of importance to society; such as generalship in the army, statesmanship, invention, authorship, sculpture, painting, music, architecture, acting."[13] Fortunately for the American taxpayer, it appears that the onset of the Depression squelched any chance that these projects would be funded.[14]

*　*　*

By 1929 both evolutionary psychology and eugenics faced grave difficulties in the scientific world, and increasingly in the popular mind. The larger historical forces that formed the bedrock of scientific culture were about to unleash a cataclysm for hereditarian explanations of human behavior. The Great Depression, which began in October of that year and continued to ground down the economy for the next decade, took its toll on science. Ideologically, the financial desperation experienced by all classes of society and the natural desire to help those in undeserved distress could not help but move Americans away from the more hard-hearted policies advocated by the extremist eugenicists such as Harry Laughlin and Madison Grant.[15] Most of the Democratic New Dealers who replaced the 1920s-era Republicans in influential political positions saw eugenics not as part of the solution, but as part of the problem.

Scientific ideology experienced a similar Leftward tilt. The "Nordics" suffered as much in the Great Depression as did everyone else; there was no solace in biologically justified paeans to the virtues of independence, hard work, and thrift. Rather, impersonal economic mechanisms and class competition seemed like much more coolly objective causes of the social phenomena witnessed in the 1930s.[16]

More prosaically, there was simply less money to go around for any scientific enterprise in the 1930s. Eugenic organizations were crushed by the sudden demise of their wealthy supporters. In 1932, the American Eugenics Society issued one of a number of urgent appeals to its erstwhile supporters for funds:

> In common with many scientific and welfare organizations [*sic!*], the American Eugenics Society has had its full share of troubles during the past few months. . . . its work is much curtailed. Some of the staff have served for months without salary and the Board has held its regular monthly meetings, well attended, and has labored hard and long in the face of many discouragements. . . . the work of our departments and committees . . . has been halted because of lack of money.[17]

Much more legitimate scientific work also was negatively impacted by the Depression. Research in comparative psychology, which had contributed so much to evolutionary psychology, was expensive. It cost much more to maintain a colony of chimpanzees, studied for their similar behavior to humans, than it did to maintain laboratory rats, which were preferred for use in simple learning experiments.

The momentous changes of the 1920s permanently altered the relative strengths of evolutionary psychology and environmental behaviorism by the end of the decade. Everyone involved in the debate clearly perceived the changing tide of battle. However, it is hard to imagine that the advocates of the biological basis of human behavior could have foreseen the fierce campaign that would be launched against them in the 1930s.

Environmental behaviorists were tolerant of the needs of the less fortunate in society, but showed much less tolerance toward those who challenged their scientific beliefs. By the 1930s, the impact of Boas and Watson could be felt throughout the entire scientific establishment. In a sense, their writ reigned supreme. There were still some holdouts from the era when evolutionary psychology was respectable, but they would either be forced into compliance or expunged from the scientific mainstream.[18]

Boas tended to categorize his colleagues as opponents or allies. As his influence grew in the anthropological establishment, many of his opponents would find that provoking his ire had undesirable consequences. The Boasians would never forgive the Smithsonian anthropologists William Holmes and Ales Hrdlicka for agreeing to serve on the National Research Council's anthropological committee, alongside the likes of such radical eugenicists as Madison Grant and Charles Davenport.[19]

In many respects, William McDougall made a particularly easy target. He had a tendency to exude a sense of his own superiority, and didn't hesitate to employ biting sarcasm in defense of his ideas. McDougall also had the unpleasant tendency to openly question his opponents' intelligence.[20]

As early as 1909, McDougall's support for instinct theory was attacked by Charles H. Judd in his presidential address to the American Psychological Association. Judd objected to McDougall's claim that human and animal behavior had the same roots. Furthermore, Judd insisted that the purpose of human behavior was not to survive environmental challenges, as McDougall alleged, but to remake the environment to suit human society.[21]

McDougall survived Judd's criticism, going on to accept an appointment at Harvard in 1920. Until the late 1920s, McDougall devoted himself to vigorously championing the still-respectable cause of the psychology of instincts. However, the harder he fought for preserving the concept of instinct, the more enemies he made. His supercilious style and relentless devotion to the concepts of instinct and vitalism (the idea that much of animal behavior was "goal-oriented") proved to be a deadly combination in the world of academic psychology. One could almost say that there was an organized hate campaign against McDougall.[22] As one psychologist recalled, an intense wave of anti-McDougall sentiment swept through the psychological community, and only died down much later.[23] Watson publicly attacked McDougall's vitalism for "returning to religion" and described his purposive psychology as "an insult to the corporate body of facts and deductions we call science."[24] To most observers, it appeared that McDougall "had been knocked out of the ring by Watson."[25] Some psychologists were shocked by a rude and insolent attack against McDougall's interest in neo-Lamarckianism by James Cattell in his address as president of the Ninth International Congress of Psychology, held at Yale in 1929.[26]

As his cause lost ground, however, McDougall turned his attention to increasingly bizarre interests. In 1928 he left Harvard for Duke University, and thereafter became engrossed in advancing the "sciences" of Lamarckianism and parapsychology.[27] McDougall himself admitted that his parapsychological research and his efforts on behalf of the notion of instincts had "destroyed my reputation as a man of science and exposed myself to the contempt of a multitude of the younger scientists of this country."[28]

McDougall's retreat seemed only to encourage his critics. In 1929, Max Schoen of the Carnegie Institute of Technology published an article entitled "Instinct and Man" in *The Scientific Monthly*. Schoen's article was a direct and withering attack against the concept of human instinct in general and against William McDougall in particular. Schoen ridiculed McDougall's definition of instinct:

William McDougall's standpoint on instinct, the influence of which has been wide and varied in education, industry and social psychology, namely, that it is "an inherited or innate psycho-physical disposition which determines the possessor to perceive, and to pay attention to, objects of a certain class, to experience an emotional excitement of a particular quality upon perceiving such an object, and to act in regard to it in a particular manner, or, at least, to experience an impulse to such action," simply states in so many complicated and mystifying words that a living organism will probably do something when confronted by a situation, in other words, that living organisms possess life. His array of specific instincts consists of an enumeration of adult activities for which no parallel whatsoever exists in the infant before the period of learning, which means the very hour of its birth.[29]

Rather, Schoen lauds the work of James Watson on infant behavior: "Certainly all the reliable evidence that we possess from Dr. Watson's experiments is against the orthodox instinct creed and in support of the conclusion that there are no instincts as such, and that 'everything we have been in the habit of calling an "instinct" to-day is a result largely of training—belonging to man's *learned behavior*.' " Schoen sharply differentiates the "instinct" of animals with the "habits" of man—these latter were merely learned conditioned reflexes. Amazingly, Schoen even denies that the infant has any "predetermined" disposition to display "curiosity, pugnacity, imitation, etc., which are usually classed as instincts." Even "mother love is no more an instinct than is love for one's country or family, a painting or a house." Working himself into a near-religious frenzy, Schoen states that "the standard criterion for instinct, namely, an act common to a species and somewhat perfect on first appearance, is inapplicable to human behavior." One wonders how Schoen explained human sexual behavior.

The above analogy to religious conviction is not gratuitous. Toward the end of his article, Schoen explicitly links his scientific and ideological convictions. The notion of instinct, that whereas it "can not be eradicated it can at least be modified and controlled," is the "psychological counterpart of the theological tenet of original sin, and like the latter, has been utilized to justify wars, industrial strife and social and political aggrandizement." Wandering even further off course, Schoen then explains to the reader that "The movement for the abolition of war is not an attempt to eliminate or suppress a pugnacity instinct by intelligence, but to change vicious and destructive social habits into constructive ones." From this, it is apparent that Schoen was quite as willing to use science to exorcise his own demons as was McDougall.[30]

McDougall also earned the opprobrium of the new generation of comparative psychologists who had for the most part accepted environmental behaviorism. Solly Zuckerman, an Oxford zoologist, was the most well-known primate researcher who opposed the contention of McDougall and

Yerkes that human behavior had an intrinsic relationship to chimpanzee behavior. In 1932, R.L. Duffus of the *New York Times* remarked that Zuckerman's work dealt a "hard blow" to those who like "to read subhuman qualities into the behavior of the human race."

> "Is this the beginning of something approaching human society?" the writer [Zuckerman] asks. "Not at all" [he responds]. "The situation is obviously one in which physiology plays the leading role, and no actions take place which require the existence of a 'social sense' to explain them." Mr. Zuckerman "tends to emphasize rather than minimize the difference between human society and that of the lower primates."[31]

The conflict between Zuckerman and McDougall came to a head at the 1934 meeting of the British Association for the Advancement of Science, in Aberdeen. The *New York Times* described the meeting as "Perhaps one of the most contentious and divisive scientific conferences ever to take place in Scotland." It focused on the issue of whether animals could think rationally. Could animal behavior be interpreted in terms of psychology, or was it simply a product of conditioned reflexes? McDougall claimed that animals were capable of a primitive form of rational behavior, whereas "Dr. Solly Zuckerman led the younger biologists in deriding the notion that animals think . . . McDougall called his opponents blind rationalists. They attacked him as a hopeless reactionary."[32]

The *New York Times* seemed to take particular delight in reveling in McDougall's decline. His later work was summed up in such headlines as "Coming Era of Vegetable Supremacy"; "Would Reform Society by Marriage Control"; "Ghosts Do Exist, McDougall Admits." Over time, such attacks wore McDougall down. He resigned from the American Psychological Association in 1934, and spent most of the last several years of his life back in England.[33]

If McDougall brought this sort of criticism on himself, the same could not be said for Edward Westermarck, the quiet Anglo-Finnish scholar interested in the anthropology of marital institutions. A series of articles published in 1931 and 1932 in the leading anthropological and sociological journals of the United States and Britain sought to destroy over four decades of accomplished work. Lord Raglan condemned Westermarck's perfectly logical thesis that humans could evolve sexual instincts that protected against deformed offspring: in this case, an aversion to incest. But humans had no instinct of any kind, not even a mating instinct, Raglan asserted. Furthermore, for some unfathomable reason, he claimed that instinct could only be a "positive," and not a "negative" force. Raglan imagined that Westermarck and "the adherents of the instinct theory of incest avoidance" required us to believe that if two males were competing for a female:

> the two males should compare pedigrees, and . . . the one more closely related to the female should instinctively renounce her. How can conduct which is inconceivable in animals be supposed to be instinctive in man? It seems to me extremely doubtful whether man has any instincts at all. I am inclined to believe that they have been completely replaced by tradition.

Raglan claimed that the incest taboo was "a purely magical idea," with no relationship to "instinct, nature, reason, nor common sense." Furthermore, the incest group was an "artificial social group," not a kin group. Primitive man's concepts of social groups did not include notions of blood relations, and so incest prohibition could not derive from an appreciation of degree of biological relatedness. Finally, incest taboos were simply too complex to be instinctual.[34]

The famous sociologist George Murdock was disturbed by Westermarck's assertion that "The marriage of mankind is not an isolated phenomenon, but has its counterpart in many animal species and is probably an inheritance from some pre-human ancestor." This premise led Westermarck to conclusions of the "utmost unreliability," Murdock claimed. Westermarck's ideas were based on the notion that there was a relationship of nature and culture; a mistaken assumption that "is now completely outmoded." Cultural phenomena were "superorganic," in a "distinct" realm independent of the laws of biology and psychology. Indeed, Murdock informed his readers that there was:

> universal agreement—if we except the extreme racialists, eugenicists, and instinctivists—that cultural behavior is socially rather than biologically determined; that it is acquired, not innate; habitual in character rather than instinctive. Culture rests, in short, not on man's specific germinal inheritance, but on his capacity to form habits under the influences of his social environment.

In a passage that must have angered Yerkes and other comparative psychologists, Murdock stated that "All analogies drawn by enthusiastic biologists between human and insect or other animal societies, fall to the ground on this point. However striking the similarities may appear, they are never more than superficial." Of course, this observation, if true, would with a single stroke destroy the basic premise of sociobiology. As evidence, Murdock pointed out that nonhuman primates were incapable of culture. Ironically enough, Murdock claimed to be enunciating these "truths" to "clear the air of dogmatism."[35]

In 1931, V.F. Calverton targeted Westermarck's belief that human monogamy and the family had parallels in other primate societies, and so was likely instinctual. These ideas were first presented in Westermarck's *History of Human Marriage*, which was regarded as the "Bible" of the social science of marriage for thirty years. However, Westermarck had "abused both evidence and observation" to derive these "absurd" conclusions, Calverton alleged. In 1927 Robert Briffault's *The Mothers* "annihilated" Westermarck's claims. Briffault demonstrated that the "family is subsumed into the group" in most primitive societies.

Given Briffault's "thoroughly valid and devastating criticisms of Westermarck's thesis," we must ask: "Why should a man's doctrine become so widely accepted when his evidences were so flimsy and fallacious? Why should his conclusion be accepted so rapidly and completely when the problems involved were so controversial? Why should he suddenly become an authority when the evidence at hand was so unauthorative?" Because, Calverton explains, such assertions as "'human marriage, in all probability, is an inheritance from some apelike progenitor'" justified conventional nineteenth-century civilization and its Victorian values.[36]

Like Westermarck, Edward A. Ross also came in for criticism in supporting the concept of human sexual instincts. In 1939 Hope Tisdale again discussed the sociology of incest. Tisdale continued the line of argument established by Lord Raglan: "We should abandon the notion that since man is a biological organism, sociology must be based on biology." For example, sociology should not attempt to understand society through reference to such supposed "biological drives" such as hunger, fear, the urge for survival, or the urge for reproduction. Further dangers lie in applying the mechanism of biological evolution to society. "Sociology has done itself a great deal of harm by trying to make the theory [of evolution] a part of itself instead of being content with watching it become a part of society." This included such now unorthodox views as focusing on the evolutionary significance of parental care of offspring.[37] The assumption that "competitiveness is innate and therefore inevitable" was also pernicious. Certainly, argued Tisdale, there was no need to invoke Edward A. Ross's claim that humans had an "instinct" against committing incest.[38]

The critiques discussed above shared an implicit assumption that human behavior was fundamentally different from that of other animals. This thesis was spelled out by the eminent British biologist and former eugenicist Julian Huxley in a 1938 article published in the *Yale Review*.[39] As we have seen, almost thirty years before Huxley had indulged in pointing out parallels between human and animal behavior, and at least hinted that these similarities were due to evolutionary links. He now renounced such speculations: humans were too complex for instincts. "Man," Huxley wrote, is "in many ways a unique animal . . . the gap between human and animal thought is much greater than is usually supposed." Unlike other animals, man's behavior "has become relatively free." This is true because man's "capacity for action" has been "largely released" from "arbitrary canalizations of instinct." In support of this conclusion, Huxley cited the unusual ability of man to ignore appearance when choosing mates.[40]

On another front, it was common among sociologists in the 1930s to oppose claims that intelligence was hereditary.[41] For example, Paul A. Witty and Harvey C. Lehman challenged this assumption in a 1931 issue of *The American Journal of Sociology*. They focused on the severely flawed (as they saw it) *Twenty-Seventh Yearbook of the National Society for the Study of Education: Nurture and Nature*. The Yearbook carried articles by such well-known educational psychologists as Edward Thorndike, who used his studies of the intelligence of siblings to substantiate the claim that intelligence was inherited.[42] Witty and Lehman were appalled at the "a priori conceptions" that informed the work published in the *Yearbook*. Not only did the *Yearbook* authors make the mistake of assuming that their intelligence tests actually measured intelligence, but "it appears that the writers who assume that intelligence is inherited in precisely the same manner as are physical traits are proceeding in direct opposition to known facts of biology" as established by John B. Watson, among others.[43]

In one interesting example of the trend, a U.S. government report on the Seventeenth International Congress of Anthropology and Prehistoric Archeology held in Bucharest, September 1–8, 1937, noted that "Dr. Ginsberg of London remarked that the authority of the modern psychological intelligence tests has now been largely discredited; so that it cannot be demonstrated that intelligent people have intelligent children, if indeed any people can be shown to possess more native mental capacity than others."[44]

Throughout the 1930s, the *New York Times* eagerly propagated the dominance of the cultural interpretation of human behavior. The *Times'* science editor, Waldemar Kaempffert, fulsomely praised those scientific researchers who accepted that human behavior was culturally conditioned. Kaempffert described Lancelot Hogben's *Genetic Principles in Medicine and Social Science* as "by far the most authoritative and judicial consideration of the social implications of genetics and eugenics that has appeared in English."

In the book, Hogben (quite rightly) condemned eugenics as a "repugnant" attempt to misuse science in an effort to justify a wide range of illegitimate prejudices. Hogben then glided from condemning eugenics to condemning geneticists who assumed that human behavior had a hereditary component. Hogben claimed that genetics had nothing to do with human behavior. Animal behaviorists were among those most guilty of such unwarranted assumptions. He criticized those who "concentrate all of their attention on the characteristics which man shares with all other animals" while neglecting "the special features which distinguish man from all other animals." For the "extreme complexity of man's social behaviour as compared with the most complicated behaviour of animals, like monkeys and dogs . . . is evident to every intelligent man or woman whose outlook has not been biased by a prolonged preoccupation with the varieties of sweet peas and mice or the pattern on the feathers of poultry."[45] Indeed, "The sneeze that accompanies the taking of snuff is almost the only example of the simple innate or unconditioned reflexes of the lower animals encountered in the social behavior of men and women."[46]

Kaempffert endorsed these conclusions. He admonished eugenicists, "who usually know nothing of genetics" to "peruse Dr. Hogben's book carefully and learn the limitations of sterilization . . . The present state of genetics and history both teach that man is a dangerous animal, but never more dangerous than when he undertakes to direct his own evolution," Kaempffert explained. [47]

Several years later, Kaempffert again praised Hogben's work in his review of J. Arthur Thomas and J.G. Crowther's collection of essays by famous environmental behaviorists, *Science for a New World*. In this instance, Hogben found that "'an infinitude of behavior patterns is consistent with the same genetic basis,' which means in plain English that society is more likely to improve by teaching its members new habits than by trying to breed a blond, blue-eyed superman. On this basis history becomes the 'descriptive study of how human-behavior patterns change from one generation to another.'"[48]

By now it will probably not surprise the reader that Waldemar Kaempffert once had been an ardent eugenicist himself. As editor of the *Popular Science Monthly*, Kaempffert enthusiastically wrote to Davenport in 1915 that "Your own subject is one which is so all-important to humanity and which must be taught by the widest kind of propaganda that you are probably in a better position to understand our problems, than any other scientist." Thus, Kaempffert intended to publish a series of articles advocating eugenics in his magazine. He wanted to "bring out the vital importance of the subject" by "discussing the incalculable social harm done by bad human protoplasm than to limit ourselves to a discussion of bad temper." A year later he reassured Davenport that "As to the subject of eugenics I need not tell you how interested I am in that."[49]

By the 1930s, others at the *New York Times* also reported favorably on the research of the environmental behaviorists. In a 1934 article, the newspaper quoted Franz Boas's authoritative statement "'behavior and not heredity is the prime factor in man's being.'"[50] Several years later, the

Times carried a summary of Clark L. Hull's presidential address to the forty-fourth meeting of the American Psychological Association. Hull sarcastically criticized those foolish enough to continue searching for complex cognitive processes:

> Because of the seemingly unique and remarkable nature of adaptive behavior, it has long been customary to attribute it to the action of a special agent or substance called "mind." But what is this mysterious thing called mind? By what principles does it operate? Are they those of the ordinary physical world or are they of the "nature of spiritual essences—of an entirely different [entity], the non-physical?" . . . he gave a demonstration with two ingenious electrical machines to show that, through application of common laws of physics [i.e., conditioning], many of the complicated mechanisms or [*sic*] human or animal behavior can be duplicated.[51]

Intelligence was a product of the environment, according to the *Times*. In 1938, it reported on the work of Beth Wellman, a researcher at the Iowa Child Welfare Research Station. Wellman found that:

> Given sufficient time and the right combination of circumstances, children will change in IQ in very large amounts. This is in essence the discovery arising from long-time studies of the same children, measured and remeasured from the pre-school ages to college. The extent of change under especially favorable circumstances may be sufficient to move a child from average intelligence to the so-called "genius" or extremely high levels. Or it may when conditions are especially unfavorable change children from average intelligence to feeble-mindedness.[52]

By the mid-1930s, evolutionarily inclined psychology professors were lamenting their great burden in turning their students away from behaviorism. William Fort Jr., a former doctoral student of William McDougall who was now a psychology professor at Mercer University, bemoaned to McDougall that "there are many students here that believe that Watson and his kind compose a group of heavenly saints whose duty is to give the only truth to mankind."[53]

If the late 1920s and early 1930s were bad for the scientific and public reception of evolutionary psychology, worse was to come. The 1930s would bring one calamity after another, which would further erode support for evolutionary psychology. Chief among them, the intertwining of American eugenics and other forms of evolutionary theories of human behavior would prove deadly.

8

The Death of Evolutionary Psychology

Man should take a "cold look" at himself, and brace himself for some unpleasant discoveries.

Niko Tinbergen

Eugenics is . . . a dangerous sword that may turn its edge against those who rely on its strength.

Franz Boas

We as eugenicists must reckon with the fact that there are many people who do not like us at all.

Samuel Jackson Holmes

End of American Eugenics

By 1929, it was obvious that American eugenics was in decline. The impact of the growing criticism of evolutionary psychology, the financial impact of the Depression, and eventually the association with Nazi eugenics would effectively destroy American eugenics within the next decade. The association of eugenics with evolutionary psychology would help drag down the latter with the former.

One by one, the eugenic organizations in the United States were inexorably decaying. Madison Grant admitted that "the future looks ominous." One major drawback was that whereas some eugenicists wanted the movement to redirect itself toward pure science, others cherished its propagandistic role. Even the Galton Society was riven by such conflict. On the one hand, Charles Davenport was distraught that the Society was being sneered at for its amateurism; on the other, Grant wanted to make sure that the eugenics movement remained "a living force in the community and not merely an obscure research into 'genes.'"[1]

The growing attacks by the environmental behaviorists also sapped the enthusiasm of the eugenicists. William McDougall's call to arms, "Are we going to take it lying down, or are we going to react to it?" failed to reignite the group's stamina.[2]

In 1929, Davenport and Samuel Holmes were both shocked to learn that the American Eugenics Society, the premier eugenics propaganda organization in the United States, had launched a new popular magazine entitled *Eugenics* without their acquiescence. They felt that the breezy, almost frivolous tone of the publication would contribute to the heap of abuse to which eugenics was now exposed. Holmes told Davenport what he already knew: "There is also a critical public of students of social science, education, and various other topics who are in various degrees hostile to eugenics, and who are looking for weak spots in its armour." *Eugenics* would only make things worse, since "it will suffer from the lack of really competent contributors and may also afford the opponents of eugenics a number of tempting opportunities for making a telling attack." Davenport wrote back that he was so incensed about being left out of the American Eugenics Society's decision to publish a popular magazine on eugenics that he was considering "severing all connections" with the Society.[3]

Due to pressure from his more sober colleagues, Davenport was distancing himself from the more radical elements of the eugenics movement by 1929. He confided in Holmes that the populists and Nordic racists threatened the very survival of serious, scientific eugenics:

> There has developed in the last few years a clear line of demarcation between human heredity and allied scientific topics, on the one hand, and a cheap sort of instruction and propaganda, on the other, which finds its lowest level, perhaps, in the advice given by the Dorothy Dixes and advice to the love lorn of the newspapers and, on the other hand, is engaged in a propaganda on the superiority of the Nordic race. As you know I have no interest in this propaganda and thoroly [*sic*] appreciated the stand of Morgan, Jennings, and Pearl, who have severed any connection with the American Eugenics Society on account of the fact that their interests are not propaganda.[4]

Davenport's new convictions were undoubtedly also a product of the Carnegie Institute's mounting concern that they were wasting money, and eroding their reputation, by continuing to endorse the "science" coming out of the Eugenics Record Office (ERO). John C. Merriam, who in 1929 was simultaneously a member of the Galton Society and president of the

Carnegie Institute of Washington, sought to quell the complaints of the growing number of critics. Merriam appointed a committee of scientists to evaluate the work of the ERO and make recommendations for its future. The committee was composed of Davenport; his assistant at the Eugenics Record Office, Harry Laughlin; a number of prominent, moderate eugenicists: Clark Wissler, Edward Thorndike, Carl Brigham, and Alfred Vincent Kidder; and Leslie Dunn, a rising star in genetics. The committee was conservative enough to refuse to make any radical changes for the time being.[5] They considered changing the name of the Eugenics Record Office to the Office of Human Genetics, but even this would not happen for another ten years.[6]

Another sign of the times was the growing tendency of the *Journal of Heredity* to distance itself from eugenics. Articles with a eugenic attitude were seen with greater rarity as the 1920s progressed. In 1930, a new editor was appointed who was more critical of eugenics than his predecessor. Thereafter, the *Journal* took a particularly cautious view of eugenics.[7]

The Third International Congress of Eugenics, held in New York City in 1932, was supposed to be a triumphal celebration of the progress of eugenics in the past decade. However, there were many reasons for the attendees to be nervous. The impact of the Depression meant less funding for the Congress, as well as a reduced attendance.[8] Charles Davenport, Harry Laughlin, Edwin Conklin, Roswell Johnson, and Paul Popenoe did manage to attend. They were dismayed when the famous geneticist Hermann J. Muller said in a speech to the delegates that the environmental influences on human behavior could not be ignored. Social inequalities, Muller said, overwhelmed any ability to differentiate human beings by genetic worth.[9] Bentley Glass claimed that Muller's speech "virtually demolished [the American Eugenics Society] . . . with a single blow. Graybreads left the hall quivering and shaken, never to recover their poise."[10] Corrado Gini also rejected the practice of American sterilization in a response to Davenport's presidential address.[11]

Roswell Johnson would also suffer declining fortunes at that time. Apparently unable to stay away from controversy for long, Johnson had managed to hurl himself to the forefront of both the birth control and eugenics movements in the 1920s. He saw both issues as intimately related. As he explained to Davenport, "One of the reasons why I have been active in the birth control movement has been . . . to try to help keep this movement as eugenic as possible." Johnson was present at Margaret Sanger's first attempt at a birth control conference, at the Town Hall Theater of Manhattan, on November 13, 1921. There he witnessed the rough police suppression of the conference. He showed up a week later, at the rescheduled meeting at the Park Theater, along with his eugenicist colleagues Clarence C. Little and Bernarr Macfadden.[12] Johnson impressed upon the delegates the need to unite birth control with eugenics. He sponsored the successful eugenic resolution to their program: "While desiring a decrease of the world birth rate in general, this Conference is well aware that this should take place on the part of individuals whose progeny would least contribute to a better race and that indeed on the part of many persons of unusual racial value that their birth rate is now too low. Therefore, be it resolved that we advocate a larger racial contribution from those who are of unusual racial value."[13] In the succeeding years, Johnson contributed to the cause of birth control and feminist hygiene by publicly speaking on behalf of the American Social Hygiene Association, an organization dedicated to reducing the spread of venereal disease.[14]

Johnson's connections with Davenport helped him climb up the hierarchy of leaders in the eugenics movement. In 1923 Johnson became a member of the Advisory Council of the Eugenics Committee of the United States. He sat on the Committee on Eugenics and Dysgenics of Birth Regulation along with Clarence Little, Edward East, Raymond Pearl, and Edwin Conklin.[15] Due to Davenport's persistent lobbying, Johnson was elected as president of the American Eugenics Society in 1926, serving for two years. In 1931, he was elected secretary/treasurer of the Society. That year he also spent two months in Japan lecturing on eugenics.[16] At other times he traveled to the Soviet Union and to Scandinavia to study eugenics programs there.[17]

Perhaps Johnson's most significant activity of 1931 was his work on birth control legislation. He spoke as a birth control advocate, along with Margaret Sanger and eugenics leader Henry Pratt Fairchild, before the birth control subcommittee of the Senate Judiciary Committee. The subcommittee was at that time collecting expert opinions on the Gillett Bill, which would have allowed licensed medical authorities to send birth control information through the mail. Johnson argued that the American upper classes were already using birth control methods, so the less fortunate should have the same option.[18] The bill, however, would fail to pass; its provisions would not become legal until 1936, as a result of the *United States vs. One Package* court decision.

Johnson created quite a few waves during his years at Pittsburgh by his promotion of eugenics and birth control. Though he attempted to reassure the administration that he eschewed communistic egalitarianism, writing the president of the University of Pittsburgh that he "probably held more extreme views on innate inequality than almost anyone else on our faculty," it appears that his activities on behalf of birth control and eugenics were not appreciated.[19]

Nor did the revision of his now "classic" work on eugenics (with Paul Popenoe), *Applied Eugenics*, have the desired impact. Much had changed since the premiere of the first edition in 1918. E.B. Reuter wrote in *The American Journal of Sociology* that the original work was "notably deficient" in its eugenic philosophy, science, and technology, and had not improved in its newest incarnation. "It is a highly uncritical piece of special pleading," Reuter concluded.[20] Capping the dramatic decline in his academic career, Johnson was dismissed from his faculty position at the University of Pittsburgh immediately after he finally completed his Ph.D. in sociology (on "International Eugenics") in 1934.[21]

Johnson resumed his wanderings, teaching at the Universities of Utah and Hawaii, and working as the executive secretary for the Palama Settlement for the poor and disadvantaged in Honolulu, Hawaii, from 1934 to 1935.[22] While in Hawaii he also worked as the executive secretary of the Hawaiian Social Hygiene Society. Johnson's activities there generated determined opposition. The head of the Society, Dr. Philip S. Platt, wrote to Paul Popenoe that Johnson faced "amazing opposition to the program," a veritable "whispering campaign against him." Johnson's efforts, Platt lamented, were "nullified, lied about and horribly misrepresented by people without any first hand contact with the man . . . The medical profession, with a few exceptions, damned it from the beginning although it never had any direct contract with the work."[23]

When funds for Johnson's position were cut, he turned for help to his friend Paul Popenoe, the director of the Institute of Family Relations. Popenoe appointed Johnson as director of the Department of Personal Service.[24] Johnson "reinvented" himself at the Institute, becoming a marriage counselor and a researcher on temperament measurement.[25] Although Popenoe was a member of the Galton Society, and the Institute had incorporated laudatory discussions of eugenics in its training and lecture materials, public opposition forced this overt eugenic content to be dropped from the Institute's programs in the mid-1930s.[26] Johnson also progressively severed his connections to eugenic organizations thereafter. His 1940 will offers an interesting illustration of his relationship to eugenics at this late date. In an earlier will he had bequeathed money to the American Eugenics Society; the new will, in order not to "embarrass" his wife, directed that the bequest be made only after her death.[27] Apparently, the eugenics community quite quickly forgot about Johnson, regardless of his former prominence and financial contributions. Johnson merited only a paragraph-long obituary in the *Eugenics Quarterly* when he died in 1967. His research work in evolutionary psychology was not mentioned.

It may be that the worsening reputation of eugenics due to the rise of Nazi Germany negatively affected Johnson's career, as it did the careers of other eugenicists and evolutionary psychologists. The Nazi Party had long championed eugenic policies, and by the early 1930s prominent German eugenicists began to endorse the party. In January 1933, Adolf Hitler became chancellor of Germany. He quickly manipulated events to establish a Nazi dictatorship. That summer, the German government announced that it would begin a massive program to sterilize eugenic "undesirables."

The Nazis eagerly mined the work of American eugenicists in constructing, and in justifying, their eugenic programs.[28] Over the course of the next several years, the scope and brutality of Germany's eugenic policies, and its relationship to the regime's ubiquitous anti-Semitism, became known in the United States. The American public by and large condemned the Nazi program. Foolishly, some of the most prominent American eugenicists rushed to defend Hitler's policies. In one instance, Paul Popenoe stated:

> Not even Hitler proposes to sterilize anyone on the ground of racial origin. My impression is that the Germans are much more anxious to weed out the undesirable elements among the non-Aryan groups. The law that has been adopted is not a half-baked and hasty improvisation of the Hitler regime, but is the product of many years of consideration by the best specialists in Germany. I must say that my impression is, from a careful following of the situation in the German scientific press, rather favorable.[29]

Hitler's policies and the lukewarm support he received from some American eugenicists doomed eugenics in the United States. As they decayed further, the eugenics societies in the United States and Britain fell under the control of moderate eugenicists who showed some sympathy toward environmental behaviorism. The transformed eugenic organizations would sever their ties to most of the few remaining evolutionary psychologists, though too late to affect the decline of either movement.

In the mid-1930s, American eugenics would become dominated by Frederick H. Osborn, a prominent Democrat and moderate eugenicist.[30] In 1928, Osborn retired young from a successful business career and decided to devote himself to eugenics and population science. His uncle, Henry Fairfield Osborn, eagerly introduced his nephew to the leading American eugenicists, and Frederick duly enrolled in the Galton Society. Osborn quickly developed an intention to dominate the American eugenics field. In 1931, Popenoe wrote to his wife that since Osborn "has ability, inclination, and means, he is likely to be one of the dominant factors in eugenics during the next quarter of a century. He and I agreed, without saying so, that we would be the controlling influence in eugenics in the future in this country."[31]

Popenoe's hubris was a bit unfounded in terms of his own importance to the eugenics movement. He was quite correct in his assessment of

Osborn's own future, however. Although initially quite impressed with the work of Laughlin and Davenport, Frederick Osborn's increasing exposure to the science of genetics and of population studies over the course of the next several years soured his perception of the current American eugenics scene. Osborn came to the conclusion that eugenics, as then practiced, had degenerated into a political propaganda movement of very dubious value, and so had lost the respect of clear-headed scientists.[32] As he would write to a correspondent decades later:

> Unfortunately the name eugenics was kept alive in the United States by a group of people who distorted Galton's ideas and purposes. They accepted the pseudo-scientific theory prevalent at that time, that the white race was superior to other races, and they gave eugenics a strong racist tinge. They under-rated the effect of bad environments in causing low I.Q.'s, delinquency and crime. They proposed sterilization for all sorts of people who in the light of today's knowledge would not be considered in any way genetically inferior. They gave eugenics a bad name, and made it easier for the public not to recognize Hitler's dreadful distortion of the word eugenics when he used it as justification for his political murders and the horror of his racial genocide. In the minds of many Americans the word eugenics still connotes racism and supermen.[33]

However, Osborn retained some hope that the concept of eugenics was salvageable if it was redirected from propaganda to cutting-edge science. Osborn wanted the "new" eugenics to accept that environmental factors played a much greater role in human behavior than the "old" eugenics was willing to admit. He also sought a more limited agenda, directed primarily to research into human genetics. He felt that only after the accumulation of a much greater degree of basic knowledge of human biology and psychology would the implementation of any eugenic political program be feasible. [34]

Laughlin worryingly described the impact of Osborn on American eugenics to his friend and protégé Domingo Ramos, a leading Cuban eugenicist:

> Osborn, while an official in the Eugenics Research Organization [sic], recently is becoming more interested in the Population Society. . . . students of eugenics cannot permit them [population scientists] to overrun the growing science of eugenics in its biological or social foundation, nor the new relationship which eugenics is making with medicine and political policies. The population group seems to be devoted to the side of sociology and environment, while the eugenics group is going in stronger all the while for basic biology in relation to practical population control.[35]

In 1935 Osborn rather reluctantly accepted the position of director of the American Eugenics Society, and added secretary/treasurer to his duties a year later. He used his influence to restructure the entire American eugenics movement. Osborn explained his plans to his friend, Franz Boas:

> A group of us have gone into the Society and have been trying to develop a sound Program which will eliminate all of the old class and race biases of eugenics, which were scientifically unjustifiable and which put the name of eugenics in a bad repute with the majority of decent people. We have made such a presentation of population, psychological and anthropological material, that the Society is fully convinced that the question of group superiorities must be entirely dropped, and proposed looking toward social changes which might bring about a more eugenic selection of births must be directed to the selection of the superior individual stocks which are to be found in every race and in every group in about equal proportions.[36]

Osborn felt that the first measure needed was a purge of the AES's leadership. Out of the 150 members of the Board of Directors and the Advisory Council, 97 were dismissed, if they hadn't quit already. Charles Davenport, Henry Goddard, Earnest Hooton, Roswell Johnson, Clarence C. Little, John C. Merriam, Henry Fairfield Osborn, Lewis Terman, Edward Thorndike, and Robert Yerkes were gone. Only Harry Laughlin, Paul Popenoe, and Frederick Osborn remained.[37]

The mass defections also spread into the Galton Society. Paleontologist William K. Gregory resigned in disgust from the Galton Society in 1935 because of its seeming support for Hitler's policies. In particular, the president of the Galton Society, Clarence Campbell, had praised Hitler and his eugenic policies at the 1935 International Congress for Population Science in Berlin. Gregory wrote to Campbell:

> On a former occasion I felt obliged to resign as chairman of the Galton Society in protest against the tendency of a number of our members to take sides with the Hitler government of Germany in its anti-Semitic and eugenical measures. The January-February number of Eugenical News, the official organ of the Galton Society, which I have just examined, affords abundant evidence that German anthropology and eugenics are still approved by leading representatives of our Society. I am convinced that the most shocking violations of elementary human rights are constantly being committed in Germany in the name of Patriotism and Science, and that the present government intends soon to demand almost everything that it believes it can take by force. Therefore, although I heartily sympathize with the efforts of our members to promote both the science of eugenics and its well-considered applications in this country, I protest again and with all my power against our Society's aligning itself with Germany. In order to make my protest the

more emphatic, I must, although with great regret, request you to submit my resignation as a Charter Fellow of the Society.[38]

Because of such losses, Davenport's retirement, and the declining health of Madison Grant, the Society closed down later that year.[39] Davenport also stepped down from the chairmanship of the National Research Council's Human Heredity Committee at about this same time. Clark Wissler and Edward Thorndike left the Committee.[40] Davenport knew that he was seeing the last days of eugenics, his life's work and passion. He admitted to Osborn (of all people), "The black buzzard of despair still seems to hang over me. 'Nothing can be done about it.' Sociology is in the saddle, and I fear to bring down the race nearly to extinction; but I suspect that the species will be able to rise again from the remnants."[41]

The 1937 meeting of the American Eugenics Society confirmed Davenport's fears. An article in the *New York Times* reported that eugenicists had finally escaped their delusions regarding the preponderant affect of genes on human behavior:

> the advocates of human betterment have given up the innocent Victorian belief in heredity as the sole agency that must be controlled if we are not to be swamped by the socially unfit. It is now recognized that undernourishment, bad housing, lack of educational and economic opportunity and a dozen environmental factors may be no longer ignored in framing eugenic policy of a democracy . . . A gospel which teaches that the right social environment is as important as the right germ plasm in scientifically improving the population will probably be more widely acceptable because it is obviously more democratic.[42]

Osborn convinced John C. Merriam, the director of the Carnegie Institute of Washington, which funded the ERO, that his new vision for eugenics was the only means for the "science" to regain its former influence.[43] In response, Merriam created yet another committee to investigate the ERO. This time, it was weighed against eugenics. Neither Davenport nor Laughlin were on the committee: Davenport because he had officially retired in 1934, Laughlin because his scientific reputation had fallen so low. Instead, the former eugenics supporter Earnest A. Hooten, along with Alfred Kidder, A.H. Schulz, and Robert Redfield, made up the committee. This time, the Committee recommended closing the ERO.[44] Merriam still dithered about, attempting to keep some sort of pseudo-eugenic research office alive. However, in 1939, Vannevar Bush, a staunch Roosevelt supporter, became president of the Carnegie Institute. Bush had no tolerance for the albatross of eugenics. He moved with all deliberate speed to close down the ERO that year and extricate the Carnegie Institute from eugenics.

By 1938, Osborn's ideas had infused all the major American eugenic organizations. With diminishing enthusiasm, he attempted to save some remnant of an American eugenic movement with a new "democratic eugenics" program. Eugenicists would henceforth be interested only in sterilizing individuals with definite, severe hereditary defects. Environmental factors would be finally given their due consideration when defining who was "defective." Also, eugenics organizations would forswear any connection between eugenics and class or race prejudices, and would refrain from endorsing political programs.

Osborn still retained some belief in heredity as a factor in determining the makeup of the human mind, especially when it came to hereditary diseases. He was well disposed to consider the danger to the human population of severe hereditary diseases. He therefore was particularly interested in the work of Dr. Franz Kallman, the foremost expert on hereditary mental disease in the United States. However, the reaction against hereditarianism was affecting even work on medical genetics.

Kallman had begun training in hereditary mental illness under the Swiss-German eugenics leader and psychiatrist Ernst Rüdin in 1934. He found that there was a roughly 15 percent incidence of schizophrenia in the parents, children, and siblings of those diagnosed with the disease. The incidence of the disease in unrelated individuals was only 1 percent. Furthermore, Kallman found that about 45 percent of identical twins had schizophrenia, whereas only 15 percent of ordinary siblings had the disease. Based on this data, Kallman concluded that schizophrenia was in part inherited. He advocated sterilization of schizophrenics as the best means to stop the spread of the disease in future populations.

Kallman did not escape the politicization of science under the Nazis. His mentor, Ernst Rüdin, was one of Germany's most important pro-Nazi scientists. Rüdin was often a spokesperson for Nazi policies at international scientific meetings, and in 1934 was serving as president of the International Federation of Eugenic Organizations. Kallman supported the Nazi line of his mentor until the 1935 Nuremberg Laws forced him to flee Germany, because Kallman was half-Jewish.

Kallman came to the United States in 1936, and found employment at the New York Psychiatric Institute. He began presenting his findings on the inheritance of schizophrenia to American audiences in 1938, giving a paper on the subject at a Eugenics Research Association meeting in early June of that year. Osborn, who heard Kallman talk, was excited about Kallman's findings. However, Osborn lamented that the prevailing environmentalist attitude among psychologists in the United States would doom any attempt Kallman might make to have his research taken seriously. As Osborn wrote to Franz Boas after the meeting:

> Dr. Kallman has available some of the best material in this country. In view of his experience in Germany he is well qualified to handle this material effectively. There is no doubt in my mind that the proposal for his research fully deserves an important place on our program. My misgivings have to do with a certain rigidity in the presentation of his conclusions which seems to me evident in his manuscript on schizophrenia and in his otherwise excellent paper presented at the Eugenics Research Association meeting last week. The group he has been working with in Germany are very much more understanding of his findings than our medical group over here, who are—many of them—exceedingly critical of hereditarian implications. For this reason, the dogmatic presentation of conclusions is, I believe, much less acceptable here than in Germany.[45]

As is apparent by now, the very word "eugenics" had become synonymous with reactionary pseudoscience, if not outright Nazism, by the end of the 1930s. The Eugenics Research Association sought to save itself by changing its name to the Association for Research in Human Heredity at the end of 1938.[46] This did little to stave off the inevitable. The organization had once over 600 members; it was now down to 205.[47]

As part of the Eugenics Research Association's transformation, it passed off the *Eugenical News* to the American Eugenics Society in 1939. Osborn had hated the old *Eugenical News*, and was determined to remake it in his image. He told his eugenicist colleague Albert Blakeslee that Grant and Laughlin had ruined the reputation of *Eugenical News* among real scientists. He had no sympathy for their propagandistic approach, "which today would be considered thoroughly unscientific." Their ideological fanaticism had "injured the scientific standing of the Record Office, and I have very direct evidence to show that it set back the scientific acceptance of eugenics in this country," Osborn explained. He fired the *Eugenical News*'s editorial board, and replaced it with individuals favoring the moderate, non-ideological eugenics he preferred.[48]

The extent to which Osborn was ready to embrace environmentalism is apparent in his March 1939 editorial statement of the new *Eugenical News*, the first issue under his direction. Osborn explained: "We are all environmentalists, and we are also all hereditarians, too, especially when we can see our way to eugenic measures which will not conflict with, but rather supplement, other social ideals." He accepted the evidence that "intelligence and socially valuable traits of personality" are "qualities which are developed by the environment acting on a complex group of genetic factors." Given the new reality, he acknowledged that "improvements must be made in the social and economic environment" before a eugenic program could be legitimately adopted in the United States. Conscious of the negative impact of Nazi eugenics on the American movement, Osborn affirmed that his eugenic program "is in accord with democratic ideals; it would supplement all other efforts at social advance; it would tend to improve the environment in which children are reared; it would leave room for many different types of human culture."[49]

Osborn's words were echoed internationally. Participants at the Seventh International Congress of Genetics, in 1939, would publish a "Geneticists' Manifesto," which reiterated that, even if it were possible and just to influence the reproductive rates of individuals with differing hereditary characteristics, it would not be valid "without economic and social conditions which provide approximately equal opportunities for all members of society instead of stratifying them from birth into classes with widely different privileges."[50] These words were sent out as World War II was breaking out in Europe; the Congress collapsed as a result of the hostilities.

Ales Hrdlicka confided his views on the state of American eugenics to Rudolf Bertheau, secretary of the American Eugenics Society, in January 1940:

> I should like to remark, however, though not necessarily for publication, that the whole field of Eugenics is not at present in a very good state. The fault lies in the fact that there have been advanced, as dogmas, various opinions and claims, before they were fully elucidated and sustained by science. The subject has become the prey of popular writers, and also of some scientific propagandists rather than researchers. It needs a lot of young blood of the best kind so that it may be reestablished as a thoroughly high-class scientific procedure.[51]

Hrdlicka's hope that "young blood" would revive American eugenics was to go unfulfilled. Several years later, Maurice Alpheus Bigelow warned Frederick Osborn that the American Eugenics Society had become an "old folks home": about 80 percent of the membership was over 60 years of age.[52] Indeed, the organization had effectively ceased to exist since the beginning of 1940.

*   *   *

Early evolutionary psychology and sociobiology died with eugenics. Environmental behaviorists had effectively destroyed the scientific reputation of evolutionary psychology and sociobiology because of their apparent association with eugenics. Now, they were even tainted with the suspicion of pro-Nazism. Responses of the evolutionary psychologists to this turn of events differed. Some denounced their former science; others offered a feeble defense. As a despondent George Estabrooks remarked to William McDougall, "The moment someone wags the finger of orthodoxy at most of us [psychologists] we promptly subside and stay in that condition."[53]

Of the various former advocates of evolutionary psychology and eugenics, Knight Dunlap was the only one who had long ago associated himself with the new behaviorist orthodoxy. By the 1930s, his earlier forays into the realm of evolutionary psychology as well as eugenics were quietly forgotten. Though twenty years before he had dreamt of a Nazi-style utopia employing forced euthanasia and sterilization, he now called eugenicists "fear propagandists" who mistakenly advocated the "nostrum" of eugenical sterilization. "The Kallikak fantasy has been laughed out of psychology, along with the even more appalling legends of the Names and the Jukes"; Dunlap averred, "but the theories involved in them still linger in popular superstitions, endorsed by many writers with other popular beliefs about heredity, and do definite damage to young persons who take the theories seriously."[54]

Ales Hrdlicka had adopted the more moderate views of Frederick Osborn. He advised Rudolf Bertheau that:

> For practical future eugenics it will be necessary to devote fully as much attention to the environment as to heredity. In other words future practical eugenics must be as much a sociological and medical as well as a biological procedure. Not until the program is thoroughly elaborated on these bases can the movement be expected to have its full, or rather near full, influence, and to be acceptable in our colleges and schools, so that it may be inculcated into the progeny, which I regard as of the foremost importance.[55]

The well-respected anthropologist Carleton Coon, though he also had strong leanings toward evolutionary psychology, was willing to accept that the groundswell of support for environmental behaviorism must have some merit. As he wrote to his cousin Carleton Putnam in 1960:

> Now about 25 years ago the scientific angle was all against you. It seemed to be proved and salted away that man is a cultural animal and there is no inheritance of instinct, intelligence, or anything else. Everything had to be learned, and he who had the best opportunity for learning came out on top.[56]

Robert Yerkes, Edward Westermarck, and Samuel Holmes were more reticent to let go of their former positions. They refused to concede much to the environmentalists, though their own influence had dramatically waned due to the changing theoretical allegiances of most American and British academics.

Yerkes wisely abandoned all work on human intelligence by the mid-1920s. He knew, as Hamilton Cravens expressed it, that "the whole field seemed too explosive a public issue and too dangerous a professional matter to pursue."[57]

Yerkes still devoted himself to his primate research center, though he found it prudent after his work on grooming to concentrate on the study of chimpanzee sensory perception, learning, and similar, less emotionally charged topics. Nevertheless, his writings occasionally hinted that his fundamental disagreement with John Watson smoldered on. In his important 1933 article on primate grooming, Yerkes alluded to the disappointment he felt over the reluctance now for his colleagues to even consider the instinctual basis for any aspect of human behavior:

> There is a strange paucity of accurate and reliable descriptions of such human activities as resemble the grooming of infrahuman primates. There is no certainty that human delousing is cultural. In early infancy as well as in primitive stages of human culture, search should be made for structurally conditioned as well as for cultural patterns of grooming.[58]

As the decade passed, Yerkes grew increasingly reflective. He became inclined to offer advice to those in the social sciences, for whatever might come of it. His views were clearly expressed in a 1937 Golden Jubilee volume of *The American Journal of Psychology.* Yerkes derided those who sought to understand animal behavior from running a thousand rats through a maze or mill "semi-mechanically and for statistical ends"; more would be leaned through the intense study of only one rat or chimpanzee. Proper study of the dominance–subordination relations in chimpanzee society in the manner Yerkes prescribed would be the most promising approach "to the comparative, genetic, and experimental study of problems of government." In concluding this retrospective work, Yerkes revealed to the reader the true reason for his twenty-year study of chimpanzee behavior. His conclusions proved that "there is obvious justification for using these animals in our search for behavioral principles and ways of producing better types of primate, including the human, and better social relations and social institutions."[59] This was a classic statement of the ultimate goals of both early-twentieth-century evolutionary psychology and eugenics, though the modern reader cannot help but to dwell on the dark potential that such social engineering suggests.

As his professional career came to a close, Yerkes became less circumspect in his public pronouncements concerning eugenics. As he wrote for a 1939 American Eugenics Society pamphlet:

> Not until we learn to control, with wisdom and foresight, both the quality and quantity of mankind shall we measurably justify our vaunted intelligence . . . It is my conviction that we should promptly avail ourselves of all reasonable eugenic and euthenic measures for racial and social betterment. Negative measures are distinctly worthwhile, since the multiplication of weakness and defects should not be permitted; positive measures are more inspiring, since their utilization requires vision and courage. Boldly and determinedly we should undertake the improvement of human quality, physically, mentally, socially, and spiritually, while endeavoring at the same time to cut off the supply of weaklings who are potential burdens to themselves and society.[60]

Yerkes also gave an opening address to the American Eugenics Society on May 6, 1941.[61]

Yerkes retired that year, passing control of the Yale Anthropoid Station (renamed the Yerkes Laboratories of Primate Biology) to his colleague, Karl Lashley.[62] Thereafter, Yerkes would involve himself in aiding the American government during the war.

Years before, Lashley had been one of John Watson's graduate students, but rebelled against Watson's rejection of the concept of consciousness.[63] Lashley could not tolerate Watson's dismissal of heredity as simply an undifferentiated base upon which learning occurs. Lashley agreed with the behaviorists (and just about everyone except McDougall) that vitalism had no role in science, but he thought it a grave error to conflate vitalism with heredity and dismiss both.[64]

Lashley's own fame was based on a 1938 paper that offered a sophisticated experimental analysis of instinctive behavior. Along much the same lines as Konrad Lorenz in Austria, Lashley argued that instinct actually referred to specific, particular neurological stimulus–response mechanisms in animals (including humans).[65]

Lashley was intensely antipathetic toward B.F. Skinner's radical behaviorism, an extension of Watson's theories, which refused to consider the internal neurological processes that underlay behavior. The British animal behaviorist W.H. Thorpe relates one particularly illuminating instance that highlights Lashley and Skinner's mutual dislike:

> When I told Skinner that we were going down to Orange Park, Florida to visit Lashley he said acidly, "Why bother to see Lashley? You won't learn anything from him!" When we got to Orange Park we soon realized that Lashley regarded the appointment of Skinner to the major chair of Psychology at Harvard as one of the greatest disasters to have befallen that University in

recent times! Lashley, though also a Professor at Harvard, went there for only two weeks a year and crammed his lectures into that space of time, "the reason for his distaste of the Harvard atmosphere being obvious!" [i.e., the Psychology Department at Harvard was dominated by Skinner and other radical behaviorists].[66]

Like Lashley, Edward Westermarck was also contending with the radical behaviorist rejection of cognition. The "new" anthropologists, Westermarck noted, objected to his approach because they believed it to be impossible to understand individual cognitive behavior. Alfred Reginald Radcliffe-Brown, president of the Anthropological Section of the British Association for the Advancement of Science, had rejected "'any explanation of a particular sociological phenomenon in terms of psychology *i.e.* of processes of individual mental activity.'" Westermarck conceded that the "evolutional anthropologists" had shown too many "methodological vagaries" in their work. But, he asked:

who can deny that even collective behaviour involves the actions of individuals? And when we speak of the customs, ideas, or religion of a people we mean by it, of course, something which the individual members of it have in common, that is, something collective. Why, them, should social anthropology refrain from studying the origins of customs and institutions?[67]

Of the early evolutionary psychologists, Samuel Holmes probably remained the most radical eugenicist. As critics tore apart eugenics in the 1930s, he refused to modify his views in the slightest. Holmes's tenacity was blasted by his opponents, while being wanly appreciated by the ever-shrinking circle of radical eugenicists. Holmes's book *The Eugenic Predicament* was reviewed in the January 1934 issue of the *Nation*. The reviewer sarcastically remarked that "An apter title for this book would be 'A Eugenicist's Predicament,' for it reflects a biologist's vacillations between his loyalty to science and his allegiance to a cult. The result is a curious hodge-podge of fact and myth, of scientifically validated conclusions and unsubstantiated prejudices." The review centered on a principle flaw in the book: the assumption that intelligence and character were hereditary and influenced one's socioeconomic position.[68] Given the frequent remarks about this review in Holmes's letters to others, he appears to have been quite taken aback by the criticism.[69]

In 1937, Holmes's dogged efforts to continue expounding eugenic ideals won him the melancholy praise of the elderly Leonard Darwin:

I look on you as one of the comparatively few authors who have not wandered away from the true eugenic path. The whole 17 years of my presidency of our society I was striving to prevent non-eugenic subjects from creeping onto our programme. Present-day topics will always oust future probabilities if they are given a chance. Hence our societies should be careful what subjects they admit to their own programmes.[70]

In 1939 Holmes was elected president of the foundering American Eugenics Society. His inaugural address, "The Opposition to Eugenics," was a frank admission that all was not well. Surveying the devastation suffered by eugenics and evolutionary psychology over the course of the past decade, Holmes found two predominant culprits: behaviorists obsessed with establishing the uniqueness of human beings, and the "intoxicated" pronouncements of some of the eugenicists themselves.

Holmes complained that the very attempt by scientists to subject the human mind to the laws of the animal realm was fraught with difficulty, given the current disposition to behaviorism. Some feared that such a devaluation of humanity's special place in the universe would deprive people of their freedom, dignity, and even their "prospects of eternal existence." Too many psychologists had latched on to the obvious role of the environment in shaping human behavior to the exclusion of any other possible influence. Many educators and social reformers were concerned that the notion that mental traits could be in part inherited challenged the worth of their professional activities. For others, the fear that radical eugenicists might succeed in promoting the sterilization (or worse) of the mentally ill led to a complete rejection of the role of inheritance in mental development or disease. Nor did the "indefensible statements" of some eugenic enthusiasts help matters.[71]

Holmes's attempt to coolly dissect the collapse of eugenics and apologize for any misunderstandings did little good for him or his cause. Furthermore, he continued promoting the eugenic movement in California, which only intensified the reaction against his activities. In the late 1930s, Holmes, Charles Matthias Goethe, Earle Ashley Walcott, and Stuart Ward conspired to create a eugenics section of the California Commonwealth Club, of which they were all members. The plan was shot down by the "Opposition," as Goethe put it. "How the enemy rushed into action and eliminated it. Censorship! And they talk of the same under Hitler, Mussolini."[72]

Later, Holmes and Goethe took up a seemingly less blatant eugenic goal: advocating special education programs for intellectually gifted children. Nevertheless, the purpose behind this endeavor was to foster a sort of mild "positive eugenics." Finding and identifying the gifted children of California, and giving them special attention in the schools, would help nurture those endowed with superior heredity. Apparently, this motive behind Holmes's and Goethe's interest in gifted education was plain to many. After Holmes's death, in 1964, Goethe encouraged Anne F. Isaacs, the executive director of the National Association of Gifted Children, to create some sort of memorial to Holmes:

> [Holmes] was really living years ahead of his time as to this whole concept of the G.C. [gifted child]. As a result, he suffered keenly in a way that perhaps no one outside his family and myself really understood. I feel it was almost Gethsemane. It seemed to be a planned attack by a certain group of aliens granted U.S.A. citizenship who seemed extremely sensitive and do not hesitate to use the "everready checkbook."[73]

Even the most determined efforts of Yerkes, Westermarck, and Holmes could not restrain the avalanche that now buried evolutionary psychology. In 1928, only about 4 percent of the members of the American Sociological Society listed "social biology" as one of their interests. In just three years, this figure dropped to less than 2.4 percent.[74] William Fielding Ogburn, in his 1934 address to the American Association for the Advancement of Science, mentioned the decline in evolutionary psychology with almost puckish understatement. "Have there been no recessions" in the "great expansion of social science?" Ogburn asked rhetorically. As hard as they were to find, there was at least one: "biological sociology." "[T]here seems to be a smaller percentage of sociologists with major interests in this field than formerly, despite a very appreciable interest in eugenics." The cause of this was not hard to find: "the natural ebb of an over-expanded interest in Darwinian biology, and the very great rise of interest in culture, which has caused considerable modification in the claims of social biology."[75]

David L. Krantz and David Allen presented a telling statistical profile of the decline of evolutionary psychology. They found that there was a "precipitous decline" in the number of articles concerning heredity in the professional journals of the 1930s and 1940s. In the 1920s, for example, the *Journal of Applied Psychology* had regularly included articles on eugenics. After 1935, they carried none.[76] Discussions of instinct theory had "almost disappeared" from psychological journals. "Even a careful experimental study, like [John Bernard] Schoolland's demonstration of innate preferences in chicks and ducklings, could go virtually unnoticed," Harold McCurdy later pointed out.[77]

One study showed that the number of articles in *Psychological Abstracts* in which the word "instinct" appeared in the title declined from 68 percent in the late 1920s to less than 40 percent by 1935. By 1955 only 8 percent of the articles used the word.[78] Psychologists were convinced that the long, bitter struggle between evolutionary psychologists and environmental behaviorists was over. The environmental behaviorists had triumphed.

Some eugenicists realized that the opprobrium that now marked eugenics had also contaminated almost the entire field of human genetics. By 1938, the wave of scientific "housecleaning" that was clearing the United States of eugenics swept into the National Research Council as well. The Human Heredity Committee, now under the chairmanship of Laurence Snyder, undertook a major self-evaluation, and was forced to accept some unpleasant facts about its condition:

> the interest of American geneticists in human genetics appears to have been waning of late, as evidenced by the almost complete absence of papers on human heredity at the various scientific meetings. This state of affairs in America, in contrast to the condition in some of the European countries, is to be deplored. It has come about, in the opinion of your committee, because of two main reasons. First, there has appeared from time to time a good deal of unscientific writing on the subject of eugenics. Since the terms "eugenics" and "human genetics" are in the minds of many persons synonymous, human genetics has suffered a loss of prestige as a result. . . . the recent attacks upon orthodox eugenics, and, indeed, upon the whole present social set-up by [Hermann J.] Muller, [Mark] Graubard and others emphasize more than ever the need for accurate facts and information on basic human genetics. . . . your Committee recommends that steps be taken to reawaken interest in human genetics on the part of investigators and laboratories.[79]

The Committee would find, however, that its quest to investigate human genetics remained tainted by the continued influence of its die-hard eugenic members. As Edmund W. Sinnott, a professor of Botany at Columbia University, judged the situation:

> [The Committee on Human Heredity] should perhaps be stimulated by the addition of new blood. . . . I take it that it has been dominated in the past rather largely by the old "war horses" of eugenics whose point of view is now somewhat discredited. The Committee should consist primarily of men with sound genetic training whose conclusions would be accepted by everyone in the field, and who have some experience of or interest in problems of human inheritance. . . . My only point is that the Committee should do some work of a sound scientific character if it is to work at all, and should not do the sort of thing, for example, that is characteristic of much of the work of the Eugenics Record Office. I take it that any geneticist would feel this way about the Committee, and that you yourself have the same attitude.[80]

Frederick Osborn's advice on the situation was sought as well. In response to their criticisms, Sewall Wright, Leslie Dunn, and William Allen were added to the Committee; Davenport and Lauglin were dropped.[81]

Notwithstanding the Committee's hopes with regard to human genetics, the field would not revive in the United States for years to come. In one illustration of this climate, an American geneticist interviewed by Daniel Kevles in 1982 recalled having been warned at the beginning of his career not to pursue his interests in human genetics. "The only thing that you can do with human genetics is develop prejudices," he was told. "And anyone who went into human genetics was immediately classified as a person of prejudice."[82]

Leslie Dunn, a geneticist and himself a recovering eugenicist, asked years later in his presidential address to the American Society for Human Genetics why the science of human genetics had been so long delayed.[83] The basic principles of genetics had been known since the rediscovery of Mendel's work at turn of the century, Dunn explained, but until the 1950s human genetics had been disappointing in the quality and quantity of research. He believed that the reason for this unfortunate lack of progress was "the long shadow" cast by eugenics "over the growth of sound knowledge of human genetics." The consequences of this blight were the deflection of attention from "the essential scientific problems" and the discouragement of "persons interested in pursuing them with human material."[84]

By the late 1930s, the closest one could get to the study of human genetics was to investigate genetics from a medical standpoint. Medical genetics was perceived as more practical than human genetics, and was directed toward understanding specific diseases. Given its more narrow focus it was possible to avoid broad or uncomfortable academic questions, though (as in Kallman's work on schizophrenia) this was not always the case.[85] During the war years, the NRC's Human Heredity Committee reoriented its funding to support research on medical genetics.[86]

Thus, even before World War II, the ideological foundations for the course of American science for the next generation of practitioners was set: human behavior was learned. Culture was an artificial human creation, founded on historical, geographical, economic, and similar factors. It was *not* a product of genetics or biology in any meaningful sense of the term. True, some psychologists (and even a small group of historians) explained human behavior using the principles of psychoanalysis, which was an explanation of human cognition. Psychoanalysis, however, sought to verify Freud's elaborate, intuitive picture of the human mind, largely separated from neurology. Freudian psychoanalysis also largely ignored the issue of human evolution.[87] It also conspicuously lacked experimental proof for its theories on the mechanism of human cognition, a reason that it quickly lost adherents after the 1950s.[88]

*   *   *

The horrors of World War II effectively buried any scientific interest that remained in eugenics or in evolutionary psychology. Realization of the full extent of the Nazi eugenic program, and its clear association with the Holocaust, seemed to suggest that it was as immoral to support evolutionary psychology as it was to support eugenics. Virtually all the remaining eugenic organizations in the United States were quietly closed down by Frederick Osborn after the war. Most of the research work on human behavior accomplished by scientists identified as eugenicists was forgotten, regardless of whether the work was eugenic in nature or not.

9

Lost in the Wilderness

Midway upon the path of our life I found myself within a dark forest, For I had lost my way.

Dante Alighieri

Most of us are aware that in a very real sense nothing can be more real than the unreal.

Ashley Montagu

Any claim that the mind has an innate organization strikes people not as a hypothesis that might be incorrect but as a thought it is immoral to think.

Steven Pinker

When ideological division replaces informed exchange, dogma is the result and education suffers.

Hunter R. Rawlings III

The first two decades after World War II were almost devoid of any meaningful research into evolutionary psychology. At the time, it would have been very unwise to have considered wandering into such an ideological minefield.[1] Most scientists depended on their professional employment for their financial security. And the Cold War atmosphere after the war did not induce many to risk their jobs and public reputation by stepping very far outside the political mainstream.

Immediately after the war, there were some last gasps of evolutionary psychology. The most important research was carried out on dog behavior at the Roscoe B. Jackson Memorial Laboratory under Clarence C. Little. Little had shown an interest in decoding dog behavior in the earliest years of his career. Little's mentor at Harvard, William Castle, mentioned Little's experiments on dogs in 1910.[2] Apparently these were not particularly productive. The subject lapsed until the summer of 1921, when Little discussed his interests with E. Carleton MacDowell and Professor Phineas W. Whiting of the University of Iowa, when all three were working at the Eugenics Record Office.[3] Davenport was sufficiently intrigued by the suggestion that he assigned the project to MacDowell. After a year MacDowell found himself frustrated and overwhelmed working with the dogs, and lacked sufficient determination to see it through. Davenport then arranged to sell the dogs to the University of Iowa, where a team composed of Whiting, Seashore, Thomas Hunt Morgan, and Samuel T. Orton set up a new dog behavior project. This approach, Whiting explained, would allow the abilities of a physiologist, psychologist, and geneticist to be focused on the project. Apparently, the project collapsed at some point.[4]

Little, however, remained interested in the potential the idea had shown. He left the Eugenics Record Office in 1922 to become president of the University of Maine, though his dedication to eugenics was undiminished. Little remained a member of the Galton Society, and served as president of the American Eugenics Society in 1929.[5] After two years at Maine, Little accepted the position of president of the University of Michigan, where he served from 1925 to 1928. He made the mistake of publicizing his eugenic views while at Michigan, inciting controversy over his presidency and forcing him to leave.[6] Little returned to research in 1929. He helped found the Roscoe B. Jackson Memorial Laboratory at Bar Harbor, Maine, to conduct research into genetics, cancer, and the breeding of research mice. He also became managing director that year of the American Society for the Control of Cancer (later known as the American Cancer Society).

Unlike some of his other eugenicist colleagues, Little refused to abandon the faith, even when it hampered his professional career. In 1932, while attending the Sixth International Congress of Genetics held at Cornell University in Ithaca, New York, Little gave a speech at the local Rotary Club where he explained to his audience:

> . . . when a sink is stopped up we shut off the faucet. We favor legislation to restrict the reproduction of the misfit. We should treat them as kindly and humanely as is possible, but we must segregate them so that they do not perpetuate their kind. Voluntary sterilization already exists in many states, but compulsory sterilization is just around the corner.[7]

Little's work with cancer research and with breeding laboratory rats kept him too busy to resume thinking about dog behavior until 1940. He approached the Rockefeller Foundation about funding the idea, but World

War II intervened before anything definite could be planned. However, in the last days of the war in Europe, Little obtained a $282,000 five-year grant from the Foundation to realize his dream.[8]

Little hired John Paul Scott as the project director. Scott had been a student of Warner Clyde Allee at Chicago. Initially interested in genetics, Scott grew frustrated at the difficulty he encountered applying his genetics studies to human problems. Thus he turned to animal social behavior as the key to understanding human society.[9]

The research project was inaugurated in 1946 with a small planning symposium on "Genetics and Social Behavior" at Jackson Laboratories in 1946. Robert Yerkes was selected to act as the general chairman of the symposium. As Scott explained to Yerkes, "The purpose of the new project is to develop fundamental basic theories underlying social phenomena. . . . any advance that can be made toward the scientific solutions of these [social] problems will bring immense benefits to mankind." Little and Scott were especially interested in learning what they could about "social control and leadership," "the causes of fighting," and the "problem of abnormal behavior" from the study. Despite promises from the acclaimed animal behaviorists Karl Lashley, Warner Cylde Allee, and Frank Beach to attend, none actually showed up for the event. The symposium attracted only twenty outside scientists.[10]

Although useful for research planning at the Jackson Laboratory, the conference did not spark a revival of interest in behavioral genetics. Scott studied dog behavior at the laboratory for the next twenty years, and wrote a number of papers that fed the revival of interest in instinct and hereditary behavior in the 1960s. Perhaps Scott's most notable find was that behavioral differences in different breeds of dogs were much more prominent when the dogs were adults than they were as pups. This finding reinforced the idea that there were genetically based behavioral instincts that were turned "on" or "off" at different stages in the life cycle.

Though Scott published about a dozen papers on his work, he created much less of a stir than had been hoped. Corrado Gini in Italy was among the few who excitedly noticed Scott's work.[11] However, overall the relatively minor interest the dog behavior studies attracted was probably something of a disappointment to the Jackson laboratories, and the project was closed down in the early 1960s.[12]

It must have been a melancholy experience for Yerkes, in retirement, to see his cherished dream of a credible evolutionary theory of human behavior lumped together with geocentrism and the humor theory of disease as just another dead end in the history of science, and seen by many as an especially pernicious one. Behaviorism and learning theory was so dominant in psychology that even comparative psychology was in its grip, and had visibly declined as a result, as Yerkes no doubt witnessed during his retirement in the 1950s.[13]

Some degree of Yerkes's anguish seems to come through in an odd little commentary he wrote for the journal *Science* in reaction to an article he had read in the British journal *Endeavor*. The *Endeavor* article, published in 1946, argued that the explanatory powers of science were limited. For example, science could not explain human aesthetics:

> the scientific method can give a great deal of information on the chemical nature of pigments, on the wavelength of the light they reflect, and on similar factors, but it is wholly unable to predict whether a picture will have an aesthetic appeal to those who see it. Nor can the scientific method be of help in those problems relating to drama, literature, and the like . . . the scientific method cannot be applied . . . to problems in which events are influenced by the philosophical values of goodness, truth, beauty, and emotions such as patriotism, fear or political conviction.

As "an American scientist who refuses to admit that it is impossible for any one person to be physicist, biologist, sociologist all in one," Yerkes was offended by the article's implication that those biologists and social scientists who apply scientific method to study man were wasting their time. He replied:

> In this emphatic statement of opinion I should wish, on the basis of my knowledge of fact, to convert all the negatives—the impossibles and improbables—into positives. Thus "inapplicable" becomes "applicable"; "it is wholly unable to predict" becomes "it is partially able to predict" . . . "qualities that cannot be measured" becomes "there are no qualities of natural phenomena which cannot be measured, although at present there are many which are not being measured or are measured very crudely and inaccurately . . ."[14]

As a further cruel irony, whereas Yerkes's own work as an early evolutionary psychologist was largely forgotten, he would be lauded as an inspiration by the three future winners of the Nobel Prize for Physiology or Medicine: Niko Tinbergen, Karl von Frisch, and Konrad Lorenz.

Tinbergen wrote to Yerkes in 1938, praising his work and asking to study at his Orange Park Primate Laboratory: "Now I should, primarily, like most to work in Your Institute, firstly because You are working with Primates . . . secondly because Your views and those of Your collaborators seem to me to be particularly stimulating."[15] Fortunately, Tinbergen was able to secure the money to realize this ambition. After his stint with

Yerkes, Tinbergen told him: "I enjoyed my stay in your laboratories. I feel I learned much about the problems and methods of American psychologists that will be of use for my future work."[16] After the War, Tinbergen could not help but lament to Yerkes about the decline of American ethology that resulted from the dominance of environmental behaviorism: "We Europeans always regret that the 'naturalistic' style of the Journal of Animal Behavior has been completely abandoned by the Journal of Comp.[i.e., Comparative] Psychology."[17]

Yerkes also had warm relations with the German ethologist Karl von Frisch. As Yerkes wrote to von Frisch: "'For several decades I have thought of von Frisch as one of the world's ablest and most productive psychobiologists. We have exchanged publications for nearly 40 years; since 1912.' I am quoting the above to remind you of the very significant role you and your work have played in my professional life as teacher and observer."[18]

Konrad Lorenz also made use of Yerkes's primate cognition work and his ideas on the evolution of human intelligence.[19]

In Italy, Corrado Gini also stubbornly kept his faith in evolutionary psychology. In his journal *Genus*, Gini hailed John Paul Scott's work with dogs for its insights into the foundations of human social behavior.[20] As he wrote in his textbook *Corso di Sociologia* in 1957, "Many institutions and customs of human society, and even some that we have been led to believe are the artificial products of civilization, are found in animal societies and probably have, at least in part, an even greater influence than we think today."[21]

The dominant line in the United States after the war is best exemplified by Ashley Montagu, one of Boas's last students.[22] Montagu continued to write a prodigious number of books into the 1970s, always backed by the same assumption: human culture was learned. Typically, in 1968 Montagu confidently asserted that "with the exception of the instinctoid reactions in infants to sudden withdrawals of support and to sudden loud noises, the human being is entirely instinctless. Those who speak of 'innate aggression' in man appear to be lacking in any understanding of the uniqueness of man's evolutionary history."[23]

These attitudes had impaired American science in favor of the Europeans. Working from their studies of the lower animals, Konrad Lorenz of Austria, Niko Tinbergen of the Netherlands, and Karl von Frisch of Germany developed new models of animal behavior (or "ethology") that reestablished the existence of instinct. They argued their case with passionate conviction, through numerous speaking tours and well-written books that slowly gained the admiration of American biologists. Yerkes's successors at his Primate Institute, Karl Lashley and Henry Wieghorst

Nissen, were among Lorenz's and Tinbergen's first American converts.[24] Momentum gradually built toward the sociobiological explosion of the 1970s.

*   *   *

By the beginning of the 1960s, interest in the concepts of instinct and hereditary behavior, which had developed in animal studies in the 1950s, was beginning to spill over into the social sciences in the United States. Those social scientists most attuned to the nuances of theoretical trends either welcomed the change as a much-needed corrective to the stifling hold that behaviorism had over their disciplines, or fretted that some new mutant variety of eugenics was again arising. To cite just one example, the president of the American Association of Physical Anthropologists, Carleton Coon, wrote to his cousin Carleton Putnam:

> The tide is turning. Heredity is coming back into fashion, but not through anthropologists. It is the zoologists, the animal behavior men, who are doing it, and the anthropologists are beginning to learn from them. It will take time, but the pendulum will swing. I stated the biological point of view at my 35th Harvard reunion last Thursday and got loud cheers and no boos. I think the Harvard Class of 1925 is a good sample of people who mould opinion.[25]

Within several years, William Hamilton would announce his concept of kin selection, and Robert Trivers would propose the theory of differential parental investment; the sciences of sociobiology and evolutionary psychology would, once again, arise.

Conclusion

Reason is the slave of the passions.

David Hume, qtd. by Edward O. Wilson,
On Human Nature, *p. 157*

After giving a lecture at Harvard in the early 1950s, British animal behaviorist W. H. Thorpe encountered Professor Edwin Boring, a leading American psychologist. Boring had noticed Wallace Craig, a former evolutionary psychologist, in the audience. Boring asked Thorpe, " 'Did you know that Wallace Craig was in the audience? I was astounded as I had supposed him dead!' " In fact, Craig had been living in Cambridge, Massachusetts for years, unnoticed.

W.H. Thorpe, The Origins and Rise of Ethology, *p. 49.*

Man is a theorizing animal. He is continually engaged in veiling the austerely beautiful outline of reality under myths and fancies of his own device. The truly scientific attitude, which no scientist can constantly preserve, is a passionate attachment to reality as such, whether it be bright or dark, mysterious or intelligible.

J.B.S. Haldane, The Causes of Evolution, *p. 170*

This book has had several goals. I wished to bring to light the extensive work that had been done earlier in the twentieth century on the sciences later called sociobiology and evolutionary psychology. A number of scientists at the time were on the right track when it came to scientific discovery, although often motivated by the wrong reasons ethically. Nevertheless, their legitimate scientific work should have been considered on its own merits, apart from the deplorable political purposes to which they may have meant it to support.

Yet, such was not to be. The other main goal of this book was to use the story of the decline and fall of evolutionary psychology as a warning of the disastrous consequences to the advancement of scientific knowledge that occurs when scientists allow ideological motives to determine their evaluation of scientific work. This warning has been sounded many times. I wish only to add another to the corpus of examples.

The nature–nurture controversy of the twentieth century, the anthropologist Derek Freeman wrote, was "an unrelenting struggle" between "two fervently held half-truths": one "overestimating biology and the other overvaluing culture." A.L. Kroeber, a student of Franz Boas, believed that evolutionists and environmentalists were separated by an "eternal chasm." This gulf was not torn open by science, but by ideological conviction.[1] In the end, the scientific reputation and work of a number of scientists in the United States and Great Britain were destroyed or forgotten, due not to their legitimate scientific work, but to their deplorable political ideologies and their eugenic propagandizing. The scientific community failed to separate science from the scientist, and would pay a heavy price in delaying the advancement of scientific knowledge.

The great irony of this story is that European scientists, approaching the matter from the perspective of ethology (animal behavior), were able to redevelop theories of instinct and other sociobiological concepts. Their freedom from connections to eugenics facilitated the acceptance of their work in the United States in the course of the 1950s and 1960s. However, much of their work had already been anticipated in the United States, in addition to other studies that were directed more precisely at understanding human behavior.

There is an abundance of evidence for this claim. Niko Tinbergen commented that he had been inspired by William McDougall's broad contextualization of animal behavior, though he was put off by McDougall's vitalism. The early field work of Lashley and Watson on bird behavior was also important.[2]

Konrad Lorenz was strongly influenced by American and British evolutionary psychologists. Among American researchers, Wallace Craig was very important for Lorenz. In many of his books, Lorenz cited Craig's work on instinct, animal learning, social and aggressive behavior, and the expression of emotions in animals. Lorenz also utilized Edward Thorndike's work on animal intelligence. Agnaldo Garcia believes that William MacDougall was "the most important British *proximate* influence on Lorenz's work." Lorenz was particularly interested in McDougall's work on instinct.[3] Not surprisingly, Lorenz was "fiercely opposed" to Watson and Skinner for their narrow focus on learning as the only tool for scientifically studying behavior. Lorenz derisively referred to behaviorism as "empty organism."[4]

Until the 1950s, however, Lorenz's research remained unknown in the United States. The British animal behaviorist W.H. Thorpe noted that in 1951, when he lectured at Harvard on the work of Tinbergen and Lorenz, he found himself amazed that "practically everything I was saying was new to my audience!" Probably one reason Lorenz's work was ignored in the United States was his known affiliation with Nazism in the late 1930s.[5] Also, by the 1950s, evolutionary psychology was once again *terra incognita* for most American psychologists.

Eckhard Hess wrote an article on "Comparative Psychology" in the 1953 *Annual Review of Psychology*. He pointed out that in the three former editions of the annual review there was "*no* mention of the work in Europe, and Lorenz's name is not mentioned, neither are the basic methods and findings of the constantly growing numbers of workers in this area of psychology [i.e., ethology]." With regard to instinct theory, Lashley had tried to bring the term back into its proper place, with no success. Hess argued that instinct was a concept critical to psychology. He concluded that "Although comparative psychology is scarcely in full flower, neither does it deserve the obituaries of recent years. Maybe current European work will turn out to be just the needed shot in the arm. Perhaps then, comparative psychology will again take its rightful place as one of the foundations on which most psychologies need to be built."[6] Ironically, Tinbergen explained that one reason he, Lorenz, and Thorpe so vigorously advocated the concept of instinct was in reaction to the sort of dismissal by American psychologists which Hess also deplored.[7]

Considering the earlier work on animal behaviorism in the United States, Thorpe thought it "extraordinary" that "the American group did not become the modern founders of ethology. They came so near it and were well in advance of workers anywhere in the world." However, further development of that science in the United States suffered an overly long "adolescence" because of the "dominance of the already established 'behaviorism,'" notwithstanding its "fatal deficiencies." In particular, Thorpe cites the work of William James and William McDougall as being "underestimated" in the United States.[8] The psychologist Harold McCurdy said much the same thing: "It is curious what antagonism McDougall's instinct doctrine once aroused in American circles. Instinct was sometimes treated as a crude superstition, and the evidence in favor of it was attacked or suppressed."[9]

In 1955, Frank Beach, one of the last surviving evolutionary psychologists of the early generation, lamented that unfortunately the "war over instincts was fought more with words and inferential reasoning than behavioral evidence."[10] Still, it would take another ten years before significant numbers of social scientists would agree with him.

Only in the mid-1960s would the American scholarly community begin a cautious and limited reassessment of the work of several of the earlier evolutionary psychologists. Donald S. Blough and Richard B. Millward, in their article "The Comparative Psychology of Learning," published in the

Annual Review of Psychology of 1965, admitted: "Learning theory, historically, guided comparative psychology into looking for what was not there, while a bit of deduction from evolutionary theory could have predicted this futility, and understandably, ethologists, until very recent times, looked upon comparative psychology as a science 'barking up the wrong tree.'"[11] The primatologist Leonard Carmichael would add, "It is now apparent that several decades of research in comparative psychology were limited in progress by the asking of inappropriate questions."[12]

Jerry Hirsh went even farther than Warren or Carmichael. In a paper delivered at a meeting of the American Psychological Association on September 4, 1966, Hirsh denounced the "stubborn and persistent opposition to the study of heredity and behavior." He attributed the relative lack of progress in behavior genetics to the unfortunate fact that "behavior geneticists have been so preoccupied with the defeat of environmentalist opposition that they have had little time for the more important task of a critical analysis of their own work." Hirsh sought the cause of this "fiasco," which "did much to steer us off the advancing stream of science": he found it in John B. Watson. Watson, Hirsh said, was so anti-evolutionary in his theories that he claimed it had been proved " 'conclusively that the vast majority of the variations of organisms are not inherited.' " "If Watson was correct," Hirsh realized, "that would have buried Darwin's theory of evolution!"[13]

Several months later, Desmond Morris claimed that the "artificial conditions of behaviour studies," such as the Skinner-box, concealed the wide range of natural behaviors in animals, who could not set out to find food, water, nest material or mates as in the natural environment. It was "extremely misleading" to overstress the value of stimulus–response studies for which behaviorism was noted. "To equate it with the whole topic of animal behaviour is like claiming that the gaming rooms of Las Vegas reflect the whole of human endeavour," Morris wryly remarked.[14]

By then there was sufficient buzz about the new research in instinct theory that the *American Sociological Review* published an article "reconsidering" the relationship between heredity and sociology.[15] The author, Bruce Eckland, directed his paper "particularly at those who believe that the ties between sociology and genetics either 'have been' or 'should be' buried."[16] Eckland lamented that "sociologists have been far more resistant to any synthesis or working arrangement between the biological and social sciences than have other investigators in these areas, including the geneticists themselves."[17] The culprit, Eckland realized, was the stranglehold that behaviorism had exercised for decades over the social sciences. He criticized environmentalist behaviorism's "stifling" effect on anthropology, which he believed had only recently begun to relax.[18] He believed that it bode ill, in the larger context of the relationship of science with society, that it had been so easy for "ideology, to influence the manner in which the scientist approaches his subject matter."[19]

As research supporting instinct theory and other evolutionary psychological concepts continued to grow, so too did the realization that a new dawn was upon the behavioral sciences. In 1968, Harold McCurdy noticed the changing attitude. He remarked that "The situation is now different. Genetics has been rediscovered, ethology has reminded us of species differences and the fascinating behavior of animals in the field, and there has been some tempering of dogmatic environmentalism by calm reflection on the realities. One begins to see McDougall's name again and some mention of his ideas with a tinge of respect. Instincts seem to be coming back into favor."[20]

R.L. Eaton would make similar observations several years later. He joined the chorus of those now chastising social scientists with their "near total obsession with learning."[21] "Even in those areas of modern psychology that seem closest to biology, for example genetic and comparative psychology, there was, until very recently, an almost absolute emphasis on learning," he noted.[22] However, "Via ethology, some of the principal questions McDougall raised are finding their way back into comparative psychology."[23]

As these realizations came to the social scientists, biologists were preparing the way for the revival of sociobiology and evolutionary psychology, as we have seen. And with this, the history of evolutionary theories of behavior comes full circle.

We can only hope that, unlike those of the past, the sound work of scientists today will not be lost due to the ideological proclivities of the future. Historians will continue to document the conflict of ideology and science in the United States and other countries, in the last century and in the current. Perhaps such work can help to keep science and ideology from once again falling into a deadly embrace.

In closing, this author wishes to invoke the words of the great biologist J.B.S. Haldane: "I would have you remember of this book only so much as I have been able to show you of the real, and forget the framework of speculation which, like myself, is transitory and ephemeral."[24]

List of Abbreviations

AHC	American Heritage Center, University of Wyoming, Laramie, Wyo.
ACS	Archivio Centrale dello Stato (State Central Archives), Rome
APS	American Philosophical Society, Philadelphia, Pa.
BLM	Bancroft Library Manuscripts, University of California, Berkeley, Calif.
CIW	Carnegie Institute of Washington Archives, Washington, D.C.
DUA	Duke University Archives, Durham, N.C.
LOC	Library of Congress Manuscript Archives, Washington, D.C.
NARA	National Archives and Records Administration, Aldephi, Md.
NAS	National Academy of Sciences Archives, Washington, D.C.
SAA	Smithsonian Anthropological Archives, Washington, D.C.
SIA	Smithsonian Institution Archives, Washington, D.C.
TSL	Truman State Library, Kirksville, Mo.
UPASC	University of Pittsburgh, Archives Service Center, Pittsburgh, Pa.
YUA	Yale University Archives, New Haven, Conn.

Notes

Introduction

1. Alcock, *The Triumph of Sociobiology*, p. 3.
2. Alcock, *The Triumph of Sociobiology*, p. 3; Pinker, *The Blank Slate*, p. 109.
3. Allen et al., "Against Sociobiology," 43. For further attacks, see Chorover, *From Genesis to Genocide*, pp. 108–10; Gould, "Biological Potential vs. Biological Determinism," p. 12+; Gould, "The Diet of Worms and the Defenestration of Prague," pp. 18–24; 64–7. Wilson presented his rebuttal in "Academic Vigilantism and the Political Significance of Sociobiology," pp. 187–90, 345. See also Wilson, *Naturalist*.
4. The term "environmental behaviorism" is used here to refer to the common philosophical underpinning of both behaviorism and Boasian anthropology: that human behavior is almost entirely molded by environment and culture, rather than instinct or heredity.
5. To give a short definition sufficient for our present purposes, sociobiology is the use of biological principles, especially evolutionary theory, to explain social behavior. Evolutionary psychology explains human psychology through the use of hypotheses that derive from our current understandings of human evolution. The difference between the two sciences is at times vanishingly small. In fact, it is sometimes impossible to distinguish them. We can think of evolutionary psychology as the attempt to understand human behavior by applying the principles of animal evolution to humans. In effect, evolutionary psychology sees human behavior as emerging from an animal foundation. Since humans are animals, their minds are the products of evolution over the course of millions of years. Through the vast majority of this time, humans and their direct ancestors lived before the advent of "civilization," in the "natural world," in which evolutionary principles guided human development. Viewed as groups (or populations) that reproduced among themselves, it would be logical to assume that any behavior or physical traits that allowed individual members of this group to reproduce (including surviving to have the chance to reproduce!) would be genetically inherited by the offspring, and allow them the same advantages. Thus, these genetic traits would become increasingly common in the group over time. Meanwhile, any other trait that also conferred reproductive or survival value would function in the same manner. One can imagine hundreds of traits, all combining in different ways and different degrees among the thousands of

individuals in a group, in a manner best described by mathematical models of genetic theory. Over time, the group as a whole would change dramatically in its behavior and appearance, or in other words, evolve. Evolutionary psychology interprets human behavior as a product of this process, modified by the incredible ability of the human brain to learn, interpret, dream, and decipher symbols. These latter influences on the human mind usually derive from the culture in which an individual lives.

6. "Catechism of the Catholic Church."
7. Bowler, *Evolution*, pp. 16–18, 21.
8. The above discussion of the philosophy of science is based on Holowchak, "Critical Reasoning & Science" (unpublished manuscript).
9. Bowler, *Evolution*, p. 191.
10. Bowler, *Evolution*, pp. 177, 299ff.
11. Badcock, *Evolutionary Psychology*, p. 73.
12. Miller, "A Review of *Sexual Selection and Human Evolution.*"
13. Cartwright, *Evolution and Human Behavior*, p. 23.
14. Bergman, "A Brief History of the Eugenics Movement"; Haller, *Eugenics*, p. 17.
15. Roll-Hansen, "Eugenics before World War II," pp. 271–2.
16. Ladd-Taylor, "Eugenics, Sterilisation and the Modern Marriage in the USA," p. 299; Dokötter, "Race Culture," pp. 467–8.
17. Spiro, "Nordic vs. anti-Nordic," pp. 41–2; Stocking, Jr., *Race, Culture, and Evolution*, p. 289.
18. Badcock, *Evolutionary Psychology*, p. 229; Cartwright, Evolution and Human Behavior, p. 23.
19. Roll-Hansen, "Eugenics before World War II," pp. 212, 216.
20. Wispé and Thompson, Jr., "The War between the Words," pp. 376–7.
21. Proctor, *Racial Science*, p. 64.
22. Roll-Hansen, "Eugenics before World War II," p. 271; Cartwright, *Evolution and Human Behavior*, pp. 20, 24.
23. Buss, *The Evolution of Desire*, p. 17.
24. Pinker, *The Blank Slate*, p. 116; Statement of Nancy Cantor on the book, *Darkness in El Dorado*, University of Michigan News Service, November 13, 2000, <http://www.ns.umich.edu/htdocs/releases/print.php?Releases/2000/Nov00/r111300a>; Tooby, "Jungle Fever."
25. Pinker, *The Blank Slate*, p. 3.
26. Pinker, *The Blank Slate*, p. 42. For some examples of neo-eugenic accusations, see Horgan, "Eugenics Revisited," pp. 122–31; "The New Social Darwinists," pp. 174–81; Allen, "Is a New Eugenics Afoot?" pp. 59–61.
27. Barkow et al. *The Adapted Mind*, pp. 5–6; Cartwright, *Evolution and Human Behavior*, p. 28.
28. Cartwright, *Evolution and Human Behavior*, p. 29; Jackson, *Physical Appearance and Gender*, p. 51.
29. The most notable exception to this last point is Thornhill and Palmer, *A Natural History of Rape*. Here, Thornhill and Palmer make recommendations for policies to decrease the incidence of rape, based on their research.
30. Degler, *In Search of Human Nature*, p. 39.

31. Proctor, *The Nazi War on Cancer*, p. 12.
32. Proctor, "The Nazi Campaign against Tobacco," p. 40.
33. Proctor, "The Nazi Campaign against Tobacco," pp. 41, 49, 50.
34. Katcher, "The Post-Repeal Eclipse in Knowledge about the Harmful Effects of Alcohol," pp. 731, 738; Pauly, "How Did the Effects of Alcohol on Reproduction Become Scientifically Uninteresting?" pp. 1, 2, 26.
35. Proctor, "The Nazi Campaign against Tobacco," p. 52.
36. Proctor, "The Nazi Campaign against Tobacco," p. 41.
37. Katcher, "The Post-Repeal Eclipse in Knowledge about the Harmful Effects of Alcohol," 729.
38. Glass, *Soldiers of God*.
39. In particular, we should applaud those who maintain that women are as intelligent as men in all areas, and are as capable in the sciences and mathematics.

1   Foundations for a "New" Synthesis

1. Tinbergen, *The Study of Instinct*; Degler, *In Search of Human Nature*, p. 223.
2. Lorenz, *On Aggression*, pp. ix, 237–8, 245–6, 271; Cohen, *Psychologists on Psychology*, p. 316; Degler, *In Search of Human Nature*, p. 228.
3. Plotkin, *Evolutionary Thought in Psychology*, p. 101.
4. Harlow, "The Nature of Love," pp. 573–685; Degler, *In Search of Human Nature*, p. 222.
5. Buss, *Evolutionary Psychology*, p. 27.
6. Robert Trivers defined "altruism" in the biological sense as "behavior that benefits another organism, not closely related, while being apparently detrimental to the organism performing the behavior, benefit and detriment being defined in terms of contribution to inclusive fitness." Trivers, *The Evolution of Reciprocal Altruism*, p. 35.
7. Hamilton, "The Genetical Evolution of Social Behaviour. I," pp. 1–2ff.; "The Genetical Evolution of Social Behaviour. II," pp. 17–18, 28–35; Buss, *Evolutionary Psychology*, p. 223.
8. Trivers, "Parental Investment and Sexual Selection," pp. 139–46; Miller, "A Review of *Sexual Selection and Human Evolution*"; Buss, *Evolutionary Psychology*, p. 42.
9. Symons, *The Evolution of Human Sexuality*, pp. 187–200.
10. Miller, "A Review of *Sexual Selection and Human Evolution*."
11. Miller, "A Review of *Sexual Selection and Human Evolution*."
12. Miller, "A Review of *Sexual Selection and Human Evolution*."
13. Fisher, *The Genetical Theory of Natural Selection*, pp. 136–7.
14. Zahavi, "Mate Selection—A Selection of Handicap," pp. 207, 213–14; Miller, "A Review of *Sexual Selection and Human Evolution*."
15. Wilson, *Sociobiology*, p. 551.
16. Wilson, *On Human Nature*, p. 32.
17. Cartwright, *Evolution and Human Behavior*, p. 183.
18. Badcock, *Evolutionary Psychology*, p. 230.

2   Recent Studies on Human Sexuality

1. Stanford, "The Cultured Ape?" p. 39.
2. See, for example, Gorman, "Sizing Up the Sexes," pp. 38–45; "Eroticism and Gender," p. 62; Allman, "The Mating Game," pp. 56–63; Cowley, "The Biology of Beauty," pp. 60–7; Begley, "You Must Remember This, a Kiss Is but a Kiss," pp. 59–63; Paul, "Blinded by Beauty," p. 17; Begley, "Boys Need Not Be Boys," p. 69; Sachs, "Dangerous Steps"; Fischman, "Why We Fall in Love," pp. 42–49; "Relationships," p. 16; Turner, "Our Cheating Hearts," p. 17; Blustain, "The New Gender Wars," pp. 42–9; Ostrow, "Do We Overestimate Our Own Desirability?," p. 24; Carpenter, "Nature's Son," pp. 70–1; Flynn, "Beauty: Babes Spot Babes," p. 10; Kluger et al., "Is God in Our Genes?," p. 62; Wright, "Dancing to Evolution's Tune," p. A11; Lemonick et al., "Honor among Beasts," p. 54; Kluger, "Ambition," p. 48; Slater, "Why We Have Children," p. 192.
3. Morris, *The Naked Ape*. Desmond Morris was a student of Niko Tinbergen. Cohen, *Psychologists on Psychology*, p. 327.
4. Miller, "A Review of *Sexual Selection and Human Evolution*"; Buss, "Sex Differences in Human Mate Preferences," pp. 1–49.
5. Jackson, *Physical Appearance and Gender*, p. 24.
6. Buss, *The Evolution of Desire*, p. 25.
7. Hawkes, "Showing Off," pp. 29–54, 32–5; Buss, *Evolutionary Psychology*, p. 78.
8. Buss, *Evolutionary Psychology*, p. 106.
9. Symons, *The Evolution of Human Sexuality*, p. 201; Buss, *Evolutionary Psychology*, p. 117.
10. Miller, "Evolution of the Human Brain through Runaway Sexual Selection"; "How Mate Choice Shaped Human Nature," pp. 87–129; Buss, *Evolutionary Psychology*, pp. 407–8; "Aesthetic Fitness," pp. 20–5.
11. Buss, *Evolutionary Psychology*, pp. 121–2.
12. Symons, *The Evolution of Human Sexuality*, pp. 185–6; Crawford et al., *Sociobiology and Psychology*, p. 341; Buss, *Evolutionary Psychology*, p. 139; Cowley, "The Biology of Beauty," p. 60.
13. Buss, *Evolutionary Psychology*, p. 139.
14. Crawford, *Sociobiology and Psychology*, p. 341.
15. Buss, *Evolutionary Psychology*, p. 118.
16. Jackson, *Physical Appearance and Gender*, p. 27.
17. Buss, *Evolutionary Psychology*, p. 110.
18. Smuts, "Sexual Competition and Mate Choice," p. 395; Buss, *Evolutionary Psychology*, pp. 116–17.
19. Cowley, "The Biology of Beauty," 60.
20. Buss, *The Evolution of Desire*, p. 110.
21. Jackson, *Physical Appearance and Gender*, p. 211.
22. Thornhill and Gangestad, "Human Facial Beauty," pp. 237–69.
23. Gangestead et al., "Facial Attractiveness," pp. 74–5, 78–82; Manning et al., "Asymmetry and the Menstrual Cycle in Women," pp. 129–30; Miller, "A Review of *Sexual Selection and Human Evolution*."

24. Cartwright, *Evolution and Human Behavior*, p. 249.
25. Jackson, *Physical Appearance and Gender*, p. 135.
26. Jones, "Sexual Selection, Physical Attractiveness, and Facial Neoteny," pp. 728, 734–5.
27. Grammer and Thornhill, "Human Facial Attractiveness and Sexual Selection," pp. 233–42; Gangestead et al., "Facial Attractiveness, Developmental Stability, and Fluctuating Asymmetry," pp. 73–85; Buss, *Evolutionary Psychology*, p. 118.
28. Smith et al., "Facial Appearance Is a Cue to Oestrogen Levels in Women," pp. 135–40; Cowley, "The Biology of Beauty," 60.
29. Cartwright, *Evolution and Human Behavior*, p. 258.
30. Buss, *The Evolution of Desire*, p. 185.
31. Jackson, *Physical Appearance and Gender*, p. 158.
32. Singh, "Adaptive Significance of Female Physical Attractiveness," pp. 293–307; Cartwright, *Evolution and Human Behavior*, pp. 243–5.
33. Zaadstra et al., "Fat and Female Fecundity," p. 484; Cowley, "The Biology of Beauty," 65.
34. Kaye et al., "The Association of Body Fat Distribution with Lifestyle and Reproductive Factors," pp. 583–91.
35. Singh, "Adaptive Significance of Female Physical Attractiveness," 293–307.
36. Jackson, *Physical Appearance and Gender*, pp. 58–60.
37. Barber, "The Evolutionary Psychology of Physical Attractiveness," p. 414; Cartwright, *Evolution and Human Behavior*, p. 256.
38. Jackson, *Physical Appearance and Gender*, p. 66.
39. Udry and Eckland, "Benefits of Being Attractive," p. 55.
40. Buss and Barnes, "Preferences in Human Mate Selection," p. 559.
41. Buss, *Evolutionary Psychology*, p. 117.
42. Nettle, "Women's Height, Reproductive Success and the Evolution of Sexual Dimorphism in Modern Humans," pp. 1919–23.
43. Buss, *The Handbook of Evolutionary Psychology*, pp. 262–3.
44. Cartwright, *Evolution and Human Behavior*, p. 258.
45. Buss, *Evolutionary Psychology*, p. 122.
46. Penton-Voak et al., "Menstrual Cycle Alters Face Preference," pp. 741–2; Fischman, "Why We Fall in Love," p. 45.
47. Schmitt, "Fundamentals of Human Mating Strategies," pp. 270, 274–5.
48. Thornhill and Palmer, *A Natural History of Rape*, p. 37.
49. Buss, *Evolutionary Psychology*, p. 118.
50. Buss and Barnes, "Preferences in Human Mate Selection," table 5, p. 567.
51. Spencer, "The Importance to Males and Females of Physical Attractiveness, Earning Potential, and Expressiveness in Initial Attraction," pp. 591–607.
52. Buss and Barnes, "Preferences in Human Mate Selection," 559–70.
53. Badcock, *Evolutionary Psychology*, p. 174.
54. Robin Baker and Bellis, *Human Sperm Competition*, pp. 1–2, 23–4, 256ff.; Gallup, Jr. and Burch, "Semen Displacement as a Sperm Competition Strategy in Humans," pp. 12–23; Buss, *Evolutionary Psychology*, p. 167.
55. Goetz and Shackelford, "Sexual Coercion and Forced In-Pair Copulation," 265; Buss et al., "Sex Differences in Jealousy," 251; Buss, *Evolutionary Psychology*, p. 167.

56. Buss et al., "Sex Differences in Jealousy," 251.
57. Buss et al., "Sex Differences in Jealousy," 252.
58. Baker and Bellis, *Human Sperm Competition*, pp. 199–202, and especially p. 200, box 8.4.
59. Walters and Crawford, "The Importance of Mate Attraction," pp. 5–30.
60. Campbell, *A Mind of Her Own*, p. 190.
61. Buss, *Evolutionary Psychology*, p. 117.
62. Buss, *Evolutionary Psychology*, p. 116.
63. Thornhill and Palmer, *A Natural History of Rape*, p. viii.
64. Alcock, *The Triumph of Sociobiology*, p. 209.
65. Thornhill and Palmer, *A Natural History of Rape*, pp. 53, 55.
66. Emlen et al., "Making Decisions in the Family," pp. 148–57; Buss, *Evolutionary Psychology*, p. 246.
67. Trivers, "The Evolution of Reciprocal Altruism," pp. 35–57; Buss, *Evolutionary Psychology*, p. 254.
68. DeVore, *Primate Behavior*, pp. 59–64; Buss, *Evolutionary Psychology*, pp. 259–60.
69. Dunbar et al., *Evolutionary Psychology*, pp. 149–65; Badcock, *Evolutionary Psychology*, pp. 262–4.

3   The Animal Nature of Humans

1. Bowler, *Evolution*, p. 289.
2. James, *The Principles of Psychology*. Cognitive psychologists today would agree with James that human beings evince instinctive fear of heights and spiders, along with snakes and the dark.
3. Cravens, *The Triumph of Evolution*, pp. 24–5.
4. Hall, "Eugenics," p. 157.
5. Hall, "Mental Science," p. 487.
6. Westermarck was a convinced eugenicist, as Karl Pearson showed in his book *The Life, Letters and Labours of Francis Galton*. Person explained that "Dr. Westermarck thoroughly approved of Galton's [eugenic] programme . . . he saw no reason why . . . [eugenic] restrictions should not be extended far beyond the existing laws of any civilized nation of to-day." At one eugenic meeting, Westermarck opined, "We cannot wait till biology has said its last word on heredity. We do not allow lunatics to walk freely about even though there may be merely a suspicion that they may be dangerous. I think that the doctor ought to have a voice in every marriage which is contracted . . . men are not generally allowed to do mischief in order to gratify their own appetites." Pearson, *The Life, Letters and Labours of Francis Galton* Vol. IIIa. *Correlation, Personal Identification and Eugenics*, p. 268.
7. Westermarck, "Methods in Social Anthropology," pp. 225–6.
8. William McDougall, *An Introduction to Social Psychology* (London: Methuen & Co., 1908), p. 351, quoted in Ellwood, *Sociology in Its Psychological Aspects*, p. 203.
9. Case, "Instinctive and Cultural Factors in Group Conflicts," p. 6.

10. McDougall, *Introduction to Social Psychology*, revised ed., pp. 27, 443.
11. Ellwood, "Obituary," p. 555.
12. Ellwood, *Sociology in Its Psychological Aspects*, p. 105; Dunlap, "The Social Need for Scientific Psychology," p. 510; Thorndike, *Elements of Psychology*, p. 15; Case, "Instinctive and Cultural Factors in Group Conflicts," 1–20. Ross, Thorndike, and Knight were all at one time or another members of the American Eugenics Society.
13. Boakes, *From Darwin to Behaviorism*, p. 206.
14. Huxley, "Introduction," in "A Discussion on Ritualization of Behaviour in Animals and Man," p. 249; Desmond Morris, introduction to Huxley, *The Courtship Habits of the Great Crested Grebe*, p. 45.
15. Huxley, *The Courtship Habits of the Great Crested Grebe*, p. 45; Huxley, "Ils n'ont que de l'âme," p. 119.
16. Huxley, *The Courtship Habits of the Great Crested Grebe*, p. 64. Although not mentioned by Huxley, similar behavior has been shown to exist in humans. See Baker and Bellis, *Human Sperm Competition*, pp. 21–2, 143 box 6.6, 214.
17. Huxley, *The Courtship Habits of the Great Crested Grebe*, p. 92.
18. Garcia, "The Psychological Literature in Konrad Lorenz's Work," pp. 105–33.
19. Craig, "Appetites and Aversions as Constituents of Instincts," pp. 104–7.
20. Craig, "Why Do Animals Fight?" pp. 264–78.
21. Trotter, *Instincts of the Herd in Peace and War*, pp. 22–3.
22. Holmes, "Darwinian Ethics," 120.
23. Wright, *The Moral Animal*, p. 67.
24. Alverdes, *Social Life in the Animal World*, p. 7.
25. Alverdes, *Social Life in the Animal World*, pp. 201–2.
26. Alverdes, *The Psychology of Animals in Relation to Human Psychology*, p. 115. Obviously, Alverdes did not realize that apparent altruism could have a self-serving role, when "altruistic" acts increased the likelihood that one's genes would be passed on to future generations indirectly, through kin.
27. Alverdes, *Social Life in the Animal World*, p. 151. Carl Groos thought along much the same lines as did Alverdes in explaining the adaptive significance of pre-adult play. See McDougall, *Introduction to Social Psychology*, revised ed., pp. 112, 115.
28. Thorpe, *The Origins and Rise of Ethology*, p. 94.
29. Darwin, *Expression of the Emotions in Man and the Animals*, p. 12; Bowler, *Evolution*, p. 289.
30. Romanes, *Mental Evolution in Man*, pp. 5–8; Bowler, *Evolution*, p. 290.
31. Ironically, the ethologists who won the Nobel Prize in 1973 would cite natural observation as the key to their work. Cohen, *Psychologists on Psychology*, pp. 318–19; Lorenz, *The Foundations of Ethology*, pp. 46–53.
32. Cravens, *The Triumph of Evolution*, pp. 10, 11, 33, 34, 39.
33. For several examples, see Ellwood, *Sociology in Its Psychological Aspects*, p. 18; McDougall, *Introduction to Social Psychology*, revised ed., p. 21; Alverdes, *Social Life in the Animal World*, p. 3; Allee, *Animal Aggregations*, pp. 3–4.
34. By studying lower animals, Lorenz, Tinbergen, and von Frisch benefited from avoiding the controversy over links between instinctual behavior in the higher

animals and human behavior. The response to Lorenz's musings on this subject in the 1960s has already been mentioned.

35. Nevertheless, the early American and English sociobiologists and evolutionary psychologists did give some quite limited attention to the behavior of lower animals, especially the social insects—subjects that would garner much greater attention in later years. See, for example, Trotter, *Instincts of the Herd in Peace and War*, pp. 20, 118; Alverdes, *Social Life in the Animal World*, pp. 3, 141–2.

36. Trotter, *Instincts of the Herd in Peace and War*, pp. 31, 32, 108–9, 111, 118, 144.

37. Alverdes, *Social Life in the Animal World*, pp. 151, 165.

38. Ridley, *The Red Queen*, p. 340.

39. Westermarck, "Methods in Social Anthropology," 224.

40. Carmichael, "Some Historical Roots of Present-Day Animal Psychology," p. 63.

41. YUA: Yerkes Papers, Box 13, Folder 230: Charles B. Davenport. Davenport to Yerkes, May 31, 1941.

42. Elliot, "Robert Mearns Yerkes," p. 487.

43. YUA: Yerkes Papers, Box 21, File 382: Henry Goddard File. Letter to Yerkes from Goddard, June 11, 1912.

44. Yerkes, "The Study of Human Behavior," pp. 625, 633.

45. Boakes, *From Darwin to Behaviorism*, p. 197.

46. YUA: Yerkes Papers, Box 5, Folder 85: E. Boring 1928–1938. Yerkes to Boring, January 13, 1932; Haslerud, preface to Yerkes, *The Mental Life of Monkeys and Apes*, p. vii.

47. Samelson, "Putting Psychology on the Map," p. 109.

48. Samelson, "Putting Psychology on the Map," p. 199; Tucker, *The Science and Politics of Racial Research*, pp. 80–1.

49. YUA: Yerkes Papers, Box 7, Folder 166: Carl Brigham. Yerkes to Brigham, October 26, 1921.

50. YUA: Yerkes Papers, Box 81, Folder 1547: Galton Society. Notice of Galton Society to Yerkes, May 17, 1919.

51. Brigham with foreword by Yerkes, *A Study of American Intelligence*; Samuelson, "Putting Psychology on the Map," p. 132.

52. TSL: Laughlin Papers. International Congresses; Eugenics: Second General Com 1920–1921. Minutes of the Meeting of the General Committee of the Second International Eugenics Congress, New York, April 10, 1920. Earnest Hooton and Raymond Pearl were also members of this committee.

53. Cravens, *The Triumph of Evolution*, pp. 181–5.

54. CIW: Robert Yerkes Files, Grants 1349 and 1480, Folder 1 of 2: Yerkes to Merriam, 22 June 1921; CIW: Robert Yerkes Files, Grants 1349 and 1480, Folder 1 of 2: Clark Wissler to Merriam, June 23, 1921.

55. Mehler, "A History of the American Eugenics Society, 1921–1940," pp. 323, 329, 401, 422, 432, 445, 448; Alroy, "Lefalophodon"; Marks, "Introduction"; Spiro, "Nordic vs. Anti-Nordic," pp. 41–2. Davenport became director of the AES in 1923, and a member of the Advisory Council in 1930.

56. APS Archives: Charles Benedict Davenport Papers, Series I. Committee on Human Behavior, October 20, 1921–October 16, 1924. Meeting of the Advisory Committee on Human Behavior at the Offices of the Carnegie

Foundation for the Advancement of Teaching, November 26, 1924. Exhibit A: Suggested outline for book on the primate collection of Madam Abreu and her experiences with monkeys and anthropoid apes.

57. Yerkes, *Almost Human*, pp. xi, 267–8.

58. Yerkes, "Mental Evolution in the Primates," p. 153.

59. The above discussion on the CIW Advisory Committee on Research on Human Behavior was reconstructed using the YUA: Robert Yerkes Papers, Box 60, Folders 1142–3: Committees CIW Advisory Committee on Research on Human Behavior 1922–1924; and the CIW Archives: File: Yerkes Grants 1349 and 1480.

60. Ironically, this volume eventually made its way to the Liberty Baptist College Library in Lynchburg, Virginia (a city with dark eugenic connotations itself). It is inscribed: "[to] Henry P. Fairchild, with the regards of Robert Yerkes, September 1925."

61. Lecture of Robert Yerkes to the Galton Society, "Recent Psycho-Biological Studies of the Chimpanzee," 65–6. Similar experiments had already been accomplished by Wolfgang Köhler during World War I. See Köhler, *The Mentality of Apes*.

62. Review of *Eugenical News, Chimpanzee Intelligence and Its Vocal Expression*, p. 124.

63. This would be called today "second order intensionality." Cartwright, *Evolution and Human Behavior*, p. 183.

64 Yerkes, "The Significance of Chimpanzee-Culture for Biological Research," pp. 71–2.

65. Review of *Eugenical News, Chimpanzee Intelligence and Its Vocal Expression*, pp. 124–5.

66. *Eugenical News* 11, 2 (February 1926): 75. This would eventually be published as Tilney and Riley, *The Brain from Ape to Man*.

67. *Eugenical News* 14, 3 (March 1929): 41–3.

68. Yerkes, "Genetic Aspects of Grooming," pp. 12–16, 21.

69. For interest in Yerkes's paper on grooming, see SAA: Ales Hrdlicka Papers, Box 69, File: Yerkes, Robert; Hrdlicka to Yerkes, February 2, 1933; Moss and Thorndike, *Comparative Psychology*.

70. APS Archives, BDd27: Charles Benedict Davenport Papers, Series I, File: Miller, Gerrit S., Jr., Davenport to Miller, April 18, 1918.

71. Miller, Jr., "The Primate Basis of Human Sexual Behavior," pp. 406–7.

72. Miller, Jr., "Some Elements of Sexual Behavior in Primates," p. 286. Miller cites Malinowski's book *Sex and Repression in Savage Society*, p. 185, as evidence of Malinowski's views on this issue. Bronislaw Malinowski was one of the foremost founders of functionalist anthropology in the 1920s. He spent much of his career conducting fieldwork in the Trobriand Islands near New Guinea, and taught at the University of London and elsewhere. Malinowski believed that all elements of culture interacted to create a stable whole. He also focused on fully explaining the complexity of a local culture *sui generis*, rather than considering one general aspect of culture (e.g., marriage) over a wide range of disparate societies.

73. Lord Raglan, "Incest and Exogamy," pp. 168, 169, 178.

74. Miller rested this assertion, rather weakly, on the fact that nonhuman primate babies could cling to their mother's fur, which was not possible with humans. Miller, "Some Elements of Sexual Behavior in Primates," 285, note 8.

75. Miller, "Some Elements of Sexual Behavior in Primates," 284. See also SI: Gerrit Miller Papers, Box 4, Folder 13: "Human Customs as a Function of Primate Behavioral Patterns."

76. These included autosexuality (i.e., masturbation), homosexuality, zoophilia, and active algolagnia. Furthermore, Miller wrote: "Harold C. Bingham, as we have seen, discovered in chimpanzees the tendencies that may easily represent the beginnings of passive algolagnia. In these directions man has done little more than specially cultivate a few elements of his primate patrimony, without producing anything especially new; though a few somewhat unimportant human behavior patterns such as enism, fetishism, kleptolagnia, narcissism, and pyrolagnia appear to stand rather specifically to his credit." Miller, "The Primate Basis of Human Sexual Behavior," 396. In his 1928 article, Miller also added exhibitionism to the list. Miller, "Some Elements of Sexual Behavior in Primates," 284, note 6.

77. Miller, "The Primate Basis of Human Sexual Behavior," 397.

78. Miller, "Some Elements of Sexual Behavior in Primates," 288.

79. Mehler, "A History of the American Eugenics Society," pp. 368–9; Holmes, "The Opposition to Eugenics," p. 351.

80. Holmes, "Darwinian Ethics," 123.

81. In a later address to the western division of the American Association for the Advancement of Science, Holmes alludes to the consternation caused in some quarters by his 1939 address. "I am somewhat shocked to find that some of my ecclesiastical misinterpreters, who seem to regard me as an apostle of iniquity, have taken me to task in a spirit that contrasts strikingly with that of the founder of the religion they profess." BLM, C-B 935: Samuel Jackson Holmes Papers, Carton 1: "Darwinianism and Ethics."

4 Earlier Studies on Human Sexuality

1. Miller, "A Review of *Sexual Selection and Human Evolution*."

2. Darwin, *The Descent of Man*, pp. 621–2.

3. Darwin, *The Descent of Man*, pp. 218–19; 221–2.

4. Geddes and Thompson, *The Evolution of Sex*, pp. 29–30.

5. Darwin gave this paper, "Mate Selection," as his presidential address to the British Eugenics Education Society on July 2, 1923.

6. Darwin, "Mate Selection," pp. 461–2; 464–5.

7. Darwin, "Mate Selection," p. 466.

8. Cravens, *The Triumph of Evolution*, p. 34.

9. Nisbet, *Marriage and Heredity*, pp. 161–3.

10. Westermarck, *The History of Human Marriage*, pp. 4, 13, 22.

11. Dunlap, *Personal Beauty and Racial Betterment*, pp. 50, 55; Holmes, *The Trend of the Race*, p. 226. Holmes reiterates this point in discussing the thought of Havelock Ellis.

12. Nisbet, *Marriage and Heredity*, p. 178.

13. Holmes, *Studies in Evolution and Eugenics*, p. 166; Thompson, "On Sexual Selection," p. 31; Carr-Saunders, *Eugenics*, p. 198.

14. Popenoe and Johnson, *Applied Eugenics*, pp. 212–13.

15. Miller, "A Review of *Sexual Selection and Human Evolution*." Huxley now held that, though human and animal behaviors might show interesting parallels, they were merely analogous, not homologous. Animal behaviors were evolved. Human behaviors were culturally determined. See Huxley, "Introduction," in "A Discussion on Ritualization of Behaviour in Animals and Man," pp. 259, 264.

16. Darwin, *The Descent of Man*, p. 605.

17. See, for example, Ellis, *Studies in the Psychology of Sex*, p. 156.

18. Williams, "Womanly Beauty among Various Races," p. 535.

19. Galton, "Composite Portraits," p. 135. Galton, *Inquiries into Human Faculty and Its Development*, pp. 230, 240–1. The Galton studies were originally mentioned in Cartwright, *Evolution and Human Behavior*, p. 249.

20. Langlois and Roggman, "Attractive Faces Are only Average," p. 118.

21. Nisbet, *Marriage and Heredity*, pp. 166, 178.

22. Dunlap, "Are There any Instincts?" p. 309. For example, at the 1931 annual meeting of the American Association for the Advancement of Science, Dunlap advocated the environmentalist position on instincts and emotions, which according to him were nothing more than cognitive myths. "Instincts are no longer believed in. They are mere words that explain nothing. Emotions are no more tangible." Kaempffert, "Year's Review of the Sciences Reveals Much New Knowledge," p. 115.

23. Dunlap, "The Social Need for Scientific Psychology," p. 515.

24. Dunlap, *Personal Beauty and Racial Betterment*, p. 66.

25. Dunlap, *Personal Beauty and Racial Betterment*, pp. 58–9.

26. Dunlap, *Personal Beauty and Racial Betterment*, pp. 30, 55.

27. Dunlap, *Personal Beauty and Racial Betterment*, p. 41.

28. Dunlap, "The Significance of Beauty," p. 209.

29. Dunlap, *Personal Beauty and Racial Betterment*, pp. 19, 21. Of course, the last remark reveals Dunlap's homophobia as well. Such remarks were hardly unprecedented among eugenicists, however.

30. Williams, "Womanly Beauty among Various Races," 533–40; Williams, "Do Men Choose Wives for Beauty?" pp. 17, 19.

31. Popenoe and Johnson, *Applied Eugenics*, p. 225.

32. Ellis, *Studies in the Psychology of Sex*, p. 195.

33. Hall, "Weighs Heart and Emotion to Tell the Scientific Secret of Love." My thanks to Mott Linn of Clark University for referring me to this article.

34. Hall, "Weighs Heart and Emotion."

35. Popenoe and Johnson, *Applied Eugenics*, p. 225.

36. Percival, "The Woman Beautiful," p. 468.

37. Percival, "The Woman Beautiful," p. 468.

38. Ellis, *Studies in the Psychology of Sex*, p. 196. Fol's study used as data photographs of 251 couples. Fol, "La resemblance entre epoux," pp. 47–9.
39. Grillenzoni, "I caratteri del fisico e del vestire come fattori demografici," p. 181.
40. Gini, "Notes," pp. 65–76.
41. McDougall, *Introduction to Social Psychology*, 14th ed., p. 279.
42. Hastings, "Dress and Its Relation to Sex," p. 122.
43. Westermarck, *The History of Human Marriage*, pp. 473, 497, 501, note 5; Ellis, *Studies in the Psychology of Sex*, p. 2 and ff.
44. Holmes, *The Trend of the Race*, p. 228; Westermarck, *The History of Human Marriage*, p. 23; Johnson, "Mate Selection," 260–1.
45. Wilkinson, "Look at Me," pp. 136–47; Todd, "Bernarr Macfadden."
46. Macfadden, *Manhood and Marriage*, pp. 30–3.
47. Westermarck, *The History of Human Marriage*, pp. 504–5.
48. Williams, "Do Men Choose Wives for Beauty?" p. 17.
49. Lenz, *Menschliche Auslese und Rassenhygiene*, p. 497. Quoted in Ludovici, *The Choice of a Mate*, p. 467, note 2.
50. Nisbet, *Marriage and Heredity*, p. 166.
51. Westermarck, *The History of Human Marriage*, pp. 5–7; Ellis, *Studies in the Psychology of Sex*, pp. 156–70; Holmes, *The Trend of the Race*, p. 226; Johnson, "Mate Selection," p. 260.
52. Ellis, *Studies in the Psychology of Sex*, p. 156.
53. "There is one general principle which applies in fixing the standards of feminine beauty in all races, namely, that the qualities most admired and desired are those which represent the secondary sex characteristics. For women, these include the broad pelvis, the greater relative size of the hips and of the gluteal region generally, the bust, the long hair, with the lack of it on the face and body, the smooth skin, the more delicate bony structure, with smaller wrists, ankles and extremities, the greater roundness of the body and of various parts of the body, the sloping shoulders, the rounded neck . . ." Furthermore, Malcolm writes, hips, buttocks, and breasts are related to successful births. Malcolm, "What Is the Ideal Female Form?" pp. 440, 442–5.
54. Williams, "Womanly Beauty among Various Races," p. 534.
55. Obituary, "Milo Hastings, 72, an Ex-Food Editor," p. 29.
56. Hastings, "The Physique Beautiful—Its Ideal Proportions In Women," pp. 6, 17, 19.
57. Malcolm, "Are Women Growing Taller?" p. 161.
58. Féré, *L'instinct sexual*, p. 44.
59. Gallichan, *The Psychology of Marriage*, p. 67.
60. Todd, "Bernarr Macfadden."
61. Hastings, "Dress and Its Relation to Sex," p. 115.
62. Hastings, "Dress and Its Relation to Sex," p. 121.
63. Gallichan, *The Psychology of Marriage*, pp. 69–70.
64. Williams, "Womanly Beauty among Various Races," p. 534.
65. Hall, *Adolescence*, p. 113; Hall, "Weighs Heart and Emotion."
66. APS Archives, BD27: Charles B. Davenport Papers, Series I, File: Johnson, Roswell H. Johnson to Davenport, March 31, 1932.

67. CIW: Roswell H. Johnson File: 1902, 1904–1906, 1919. Davenport to Woodward, March 2, 1905.

68. APS Archives, BD27: Charles B. Davenport Papers, Series I, File: Johnson, Roswell H. Johnson to Davenport, August 20, 1897.

69. APS Archives, BD27: Charles B. Davenport Papers, Series I, File: Johnson, Roswell H. Johnson to Davenport, June 3, 1903.

70. APS Archives, BD27: Charles B. Davenport Papers, Series I, File: Johnson, Roswell H. Johnson to Davenport, June 3, 1903.

71. CIW: Roswell H. Johnson File: 1902, 1904–1906, 1919. Johnson to the Secretary of the CIW, August 21, 1904.

72. CIW: Roswell H. Johnson File: 1902, 1904–1906, 1919. Outline of a proposed Vivarium for Experimental Evolution, July 3, 1902.

73. Johnson, "The Evolution of Man and Its Control," pp. 53, 54, 55, 58.

74. APS Archives, BD27: Charles B. Davenport Papers, Series I, File: Johnson, Roswell H. Johnson to Davenport, August 13, 1910.

75. APS Archives, BD27: Charles B. Davenport Papers, Series I, File: Johnson, Roswell H. Johnson to Davenport, October 7, 1910.

76. Jacques Cattell, *American Men of Science*; Marquis, *Who's Who in America 1928–1929*, p. 1155; CIW: Roswell H. Johnson File: 1902, 1904–1906, 1919. R.S. Woodward to Johnson, January 20, 1906; CIW: Roswell H. Johnson File: 1902, 1904–1906, 1919. Application for a Grant in Aid of Research, May 17, 1905; CIW: Roswell H. Johnson File: 1902, 1904–1906, 1919. Application for a Grant in Aid of Research, February 6, 1903.

77. APS Archives, BD27: Charles B. Davenport Papers, Series I, File: Johnson, Roswell H. Johnson to Davenport, November 25, 1913.

78. Johnson, "The Improvement of Sexual Selection," p. 531. It was not unreasonable for Gilmore to assume that most thirty-year-old American women before World War I already would have married, if they were ever to do so. Further information about this study also comes from Popenoe and Johnson, *Applied Eugenics*, p. 217. I have been unable to locate the original study.

79. Johnson, "The Improvement of Sexual Selection," p. 531; Popenoe and Johnson, *Applied Eugenics*, p. 217; Carr-Saunders, *Eugenics*, p. 198; Naly, "Facial Appearance and other Factors in Mate Selection among College Graduates," p. 8; Grillenzoni, "I caratteri del fisico e del vestire come fattori demografici," pp. 131–2; Holmes and Hatch, "Personal Appearance as Related to Scholastic Records and Marriage Election in College Women," p. 74; Gini, "Notes," pp. 65–76. Since Ronald A. Fisher and Julian Huxley read Carr-Saunder's manuscript, and J. Arthur Thompson was on the series editorial board for which the book was produced, they also must have been aware of the Gilmore study.

80. Naly, "Facial Appearance and other Factors in Mate Selection among College Graduates," p. 9.

81. Naly is listed as a "subscribing member" of the AES in the August 1930 "List of Members of the American Eugenics Society," in LOC: Margaret Sanger Papers, container 63, reel 41. My thanks to Patrick Kerwin for this information.

82. Naly, "Facial Appearance and other Factors in Mate Selection among College Graduates," p. 8. Whiting's study was poorly conceived. For example, the

judges were asked to make qualitative, rather than quantitative, evaluations of the women's attractiveness. Because of this, it was not possible to compare results between judges, and so made meaningful statistical analysis almost impossible. Whiting was teaching at Catawba College in Salisbury, North Carolina, when she began her research into human sexual selection. Her husband, Phineas Whiting, was a professor at the University of Pittsburgh, which might well explain how Naly encountered Whiting's research. APS Archives, BD27: Charles B. Davenport Papers, Series I, File: Whiting, Anna R. Whiting to Davenport, February 12, 1928.

83. Naly, "Facial Appearance and other Factors in Mate Selection among College Graduates," Table I Marriage and Selection, p. 11.

84. Naly, "Facial Appearance and other Factors in Mate Selection among College Graduates," Graph I Marriage-Selection, p. 14b.

85. Naly, "Facial Appearance and other Factors in Mate Selection among College Graduates," Graph II "Marriage Rate and Total Score for Appearance. Classes of 1910, 1911, 1913 Smith," p. 14a.

86. Elsewhere in the study, the total score for appearance was plotted to show whether there was any relation between attractiveness and the number of years elapsing between graduation and marriage. "A very jogged and irregular curve resulted, showing that there was no relation between facial appearance and early marriage so far as these three college classes were concerned. The more attractive and more beautiful did not marry any earlier than any other group." Examining the variability of the judges' rankings for each subject, Naly concluded, "The judges vary considerably in agreement, but not out of reason." Elsewhere she said that the judges agreed "fairly well" in their rankings of each subject. Naly, "Facial appearance and other factors in mate selection among college graduates," pp. 13, 21.

87. For more on Latin eugenics, see Gillette, *The History of Latin Eugenics* (manuscript).

88. Israel and Nastasi, Scienza e razza nell'Italia fascista, p. 114.

89. De Grazia, *How Fascism Ruled Women*, pp. 48–9.

90. Relatively older women were chosen so that their life fertility rates could be calculated.

91. Grillenzoni, "I caratteri del fisico e del vestire come fattori demografici," p. 136.

92. Grillenzoni, "I caratteri del fisico e del vestire come fattori demografici," Table II, p.o., p. 149.

93. Grillenzoni, "I caratteri del fisico e del vestire come fattori demografici," Table II, m'.v., p. 150.

94. Grillenzoni, "I caratteri del fisico e del vestire come fattori demografici," Table VII, m.o., p. 153.

95. Grillenzoni, "I caratteri del fisico e del vestire come fattori demografici," Table VII, m'.v., p. 153.

96. Grillenzoni, "I caratteri del fisico e del vestire come fattori demografici," Table VII p.o., p. 152.

97. Grillenzoni, "I caratteri del fisico e del vestire come fattori demografici," pp. 151, 157.

98. Grillenzoni, "I caratteri del fisico e del vestire come fattori demografici," Table III, p.o., p. 158.

99. Grillenzoni, "I caratteri del fisico e del vestire come fattori demografici," Table VIII, m, p. 163.

100. Grillenzoni, "I caratteri del fisico e del vestire come fattori demografici," p. 161.

101. Grillenzoni, "I caratteri del fisico e del vestire come fattori demografici," Table III, m., p. 161.

102. Grillenzoni, "I caratteri del fisico e del vestire come fattori demografici," Table XIII, m., p. 164.

103. Grillenzoni, "I caratteri del fisico e del vestire come fattori demografici," pp. 163–4.

104. Grillenzoni, "I caratteri del fisico e del vestire come fattori demografici," Table XVII, m.v., p. 183.

105. Horn, "Constructing the Sterile City," p. 586. De Grazia, *How Fascism Ruled Women*, pp. 212–13, 296–7.

106. Grillenzoni, "I caratteri del fisico e del vestire come fattori demografici," p. 132.

107. Gini, "Notes," pp. 65–76.

108. "Samuel Holmes, Zoologist, Dead," p. 87.

109. Eakin et al., "1965, University of California."

110. Holmes and Hatch, "Personal Appearance as Related to Scholastic Records," p. 65.

111. Holmes and Hatch, "Personal Appearance as Related to Scholastic Records," p. 68.

112. Holmes and Hatch, "Personal Appearance as Related to Scholastic Records," p. 74.

113. Holmes and Hatch, "Personal Appearance as Related to Scholastic Records," pp. 74, 76.

114. "An Ideal Wife's Qualities," p. 60. The magazine defined the study's terms as follows: "Health covers wholesomeness, physical efficiency, fitness for maternity, love of outdoor life, and so forth." It constitutes the "physical assets" of the ideal wife. In regard to beauty: "One-half of those who mention it, speak of beauty of face and figure—the form receiving the larger proportion of votes." Disposition was defined as: "Loving, affectionate, sympathetic, sweet tempered, tactful, cheerful, good-natured, warm hearted, sociable." Most men insisted that their wives' dress "should be 'neat and simple and beautiful always.'" The data was based on the responses to the survey of over one thousand readers (pp. 60–1).

115. Holmes, *The Trend of the Race*, pp. 226–7.

116. Holmes, *The Trend of the Race*, pp. 226–7.

117. Darwin, "Mate Selection," pp. 463–4. The study is discussed in Popenoe and Johnson, *Applied Eugenics*, p. 219.

118. Harrison Hunt, "Matrimonial Views of University Students," pp. 14–15.

119. Hunt, "Matrimonial Views of University Students," 19–20.

120. Hunt, "Matrimonial Views of University Students," p. 21.

121. Hunt, "Matrimonial Views of University Students," p. 21.

122. See Hunt, *Some Biological Aspects of War*; Michigan State University Archives and Historical Collections, http://www.si.umich.edu/HCHS/HCHS-GUIDE/hchs.source563.html, (accessed June 20, 2004).
123. Popenoe, *Modern Marriage*, p. 36.
124. Popenoe, *Modern Marriage*, p. 36.
125. Holmes, *The Trend of the Race*, p. 226. Holmes quotes Howard, *History of Matrimonial Institutions*, p, 216, and Westermarck, *History of Human Marriage*, p. 212, as evidence.
126. Nisbet, *Marriage and Heredity*, p. 169.
127. Ferrero, "The Problem of Women from a Biosocial Point of View," pp. 263, 265, 271.
128. Westermarck, *The History of Human Marriage*, p. 212. This passage was also summarized by Holmes in *The Trend of the Race*, p. 226.
129. Holmes's expectations have been confirmed by David Perrett and others. See, for example, Penton-Voak et al., "Menstrual Cycle Alters Face Preference," 741–2.
130. Holmes, *The Trend of the Race*, pp. 223, 226.
131. Johnson, "Mate Selection," p. 260.
132. Dunlap, *Personal Beauty and Racial Betterment*, p. 22.
133. Dunlap, "The Significance of Beauty," p. 209. Compare this to Buss, *Evolutionary Psychology*, p. 117.
134. Dunlap, *Personal Beauty and Racial Betterment*, p. 91.
135. Gallichan, *The Psychology of Marriage*, p. 70.
136. Nisbet, *Marriage and Heredity*, p. 169.
137. "The Ideal Husband," p. 42.
138. Hall, "Weighs Heart and Emotion."
139. "The Ideal Husband," pp. 38–43. Slightly less than one thousand respondents participated in this survey. The magazine's discussion of its study also remarks on the importance of these characteristics for proper eugenic selection of spouses (pp. 38–9, 43).
140. Popenoe and Johnson, *Applied Eugenics*, p. 219.
141. Popenoe, *Modern Marriage*, p. 83.
142. Popenoe, *Modern Marriage*, pp. 83–4.
143. Ellwood, *Sociology in Its Psychological Aspects*, p. 235.
144. Holmes, *The Trend of the Race*, p. 223.
145. Darwin, *The Descent of Man*, p. 617.
146. Miller, "Some Elements of Sexual Behavior in Primates," p. 289.
147. Miller, "The Primate Basis of Human Sexual Behavior," p. 401; Miller, "Some Elements of Sexual Behavior in Primates," p. 290.
148. Groos, *Play of Animals*, pp. 242–3; in Thompson, "On Sexual Selection," p. 28.
149. McDougall, *Introduction to Social Psychology*, 14th ed., p. 277.
150. Baker and Bellis, *Human Sperm Competition*, pp. 8–11, 13.
151. Darwin, *The Descent of Man*, p. 619.
152. Hall, *Adolescence*, pp. 111–12.
153. Johnson, "Mate Selection," p. 263.
154. Campbell, *A Mind of Her Own*, p. 188.

155. Hall, *Adolescence*, pp. 357–8.
156. "The Ideal Husband," p. 41.
157. Hall, *Adolescence*, pp. 111–12; Hall, "Weighs Heart and Emotion."
158. Westermarck, *The History of Human Marriage*, pp. 462–3, 471.
159. Westermarck, *The History of Human Marriage*, p. 301.
160. Gesell, "Jealousy," p. 447, quoted in Ellis, *Studies in the Psychology of Sex*, p. 565.
161. McDougall, *Introduction to Social Psychology*, 14th ed., p. 65.
162. See Lang and Atkinson, *Social Origins*.
163. McDougall, *Introduction to Social Psychology*, 14th ed., pp. 194–7.
164. Westermarck, *The History of Human Marriage*, pp. 1–2, 23–4, 256ff.

5   Evolution, Ethics, and Culture

1. Darwin, *The Descent of Man*, Vol. II, pp. 98, 107.
2. Trotter, *Instincts of the Herd in Peace and War*, p. 24; Holmes, "Darwinian Ethics," p. 120.
3. Trotter, *Instincts of the Herd in Peace and War*, p. 29.
4. Allee, *Animal Aggregations*, p. 361.
5. Alverdes, *Social Life in the Animal World*, p. 174.
6. Yerkes, "Mental Evolution in the Primates," p. 137.
7. A good discussion of current research on nonhuman primate altruism can be found in de Waal, *Our Inner Ape*.
8. Yerkes, "Genetic Aspects of Grooming," p. 18.
9. Yerkes, "Genetic Aspects of Grooming," pp. 18–19; see also Yerkes, "Primate Cooperation and Intelligence," p. 267.
10. Nissen and Crawford, "A Preliminary Study of Food-Sharing Behavior," pp. 383–419, 415.
11. Nissen and Crawford, "A Preliminary Study of Food-Sharing Behavior," p. 415.
12. Nissen and Crawford, "A Preliminary Study of Food-Sharing Behavior," p. 401.
13. Nissen and Crawford, "A Preliminary Study of Food-Sharing Behavior," p. 414.
14. Nissen and Crawford, "A Preliminary Study of Food-Sharing Behavior," p. 415; M.P. Crawford, "Cooperative Behavior in Chimpanzee," read at the Ann Arbor meeting of the American Psychological Association, September 5, 1935.
15. Nissen and Crawford, "A Preliminary Study of Food-Sharing Behavior," p. 415.
16. Ross must also rank as one of the most racist of American sociologists. In his book, *The Old World in the New*, he complained that many recent immigrants "even when they were 'washed, combed, and in their Sunday best,'" were "hirsute, low browed, big faced persons of obviously low mentality," and were totally out of place in civilized clothes, "since clearly they belong in skins in wattled huts at the close of the Great Ice Age." To his "practiced" eye, "the physiognomy of certain groups unmistakably proclaims inferiority of type. I have seen gatherings of the foreign-born in which narrow and sloping foreheads were the rule. The shortness and smallness of the crania were very noticeable. There was much facial asymmetry . . . . In every face there was something—wrong-lips thick, mouth coarse, upper lips too long, cheek bones too high,

chin poorly formed, the bridge of the nose hollowed, the base of the nose tilted, or else the whole face prognathous. There were so many sugar-loaf heads, moon-faces, slit mouths, lantern-jaws, and goose-bill noses that one might imagine a malicious jinn had amused himself by casting human beings in a set of skew-molds discarded by the creator." Ross, *The Old World in the New*, pp. 285–6; Tucker, *The Science and Politics of Racial Research*, p. 74.

17. Ross, *Social Control*, p. 7.
18. Ellwood, *Sociology in Its Psychological Aspects*, pp. 216, 315.
19. Holmes, "Evolution and the Ethical Ideal," p. 10.
20. Johnson, "Mate Selection," p. 261.
21. Haldane, *The Causes of Evolution*, p. 131.
22. Holmes, "Darwinian Ethics," p. 123. Holmes would later develop this thesis into a book after World War II: *Life and Morals*. Robert Yerkes would read this work, and heap fulsome praise on Holmes for writing it: "I . . . give you my best thanks for a service to mankind. It is needless to assure you that I am in complete agreement with your major assumptions and the views which you develop and defend in your admirably clear and brief presentation of an incomparably important subject." BLM, BANC MSS C-B 935: Holmes, Samuel J., Box 2, Correspondence I-Z, File: R.M. Yerkes, Robert M. Yerkes to Holmes, January 27, 1949. Years later, Holmes would blame "Catholic censorship" for the poor sales the book would garner. BLM, BANC MSS C-B 935: Holmes, Samuel J., Box 2, Correspondence I-Z, File: Outgoing letters, SJH, 1924–1961. Holmes to The Macmillian Company, June 21, 1960.
23. Miller, "Some Elements of Sexual Behavior in Primates," p. 286; Miller, "The Primate Basis of Human Sexual Behavior," pp. 393, 395, 403. For Bingham's observation, see Bingham, *Sex Development in Apes*, p. 92.
24. Johnson, "The Improvement of Sexual Selection," p. 515; Holmes, *The Trend of the Race*, pp. 225–6.
25. Ellwood, *Sociology in Its Psychological Aspects*, p. 229.
26. Johnson, "The Improvement of Sexual Selection," p. 516; Johnson, "The Evolution of Man and Its Control," p. 60.
27. Holmes, "The Role of Sex in the Evolution of the Mind," p. 202.
28. McDougall, *Introduction to Social Psychology*, 14th ed., pp. 269–70.
29. Miller, "Evolution of Human Music through Sexual Selection," pp. 271–300; Miller *The Mating Mind*, pp. 20–5.
30. Alverdes, *Social Life in the Animal World*, pp. 204–5; Johnson, "The Evolution of Man and Its Control," p. 60; Ellwood, *Sociology in Its Psychological Aspects*, p. 214.
31. BLM, C-B 935: Samuel Jackson Holmes Papers, Carton 1: "The Ethics of Enmity (The Natural History of Enmity); (Group Antagonism and Mutual Aid)," unpublished manuscript, ca. 1945.

6 The Rise of Environmental Behaviorism

1. The quotation continues: "The interpretation of what happens in the case of man or animals is thus a matter of the psychological structure of the observer;

the psychical quality of the individual scientist is responsible for the nature of the surrounding world which he builds up for himself; it decides what things and processes are invested with actuality in it, and what are ignored as unimportant. Furthermore, every variety of world view can be linked with the results of our observation in nature." Alverdes, *The Psychology of Animals in Relation to Human Psychology*, p. 139.

2. Cravens, *The Triumph of Evolution*, pp. 10–11.
3. APS Archives, BD27: Charles Benedict Davenport Papers, Series I, File: Campbell, C. G. (Clarence Gorden), 1868–1956. Campbell to Davenport, June 10, 1928.
4. APS Archives, BD27: Charles Benedict Davenport Papers, Series I, File: Campbell, C. G. (Clarence Gorden), 1868–1956. Campbell to L[orande] L[oss] Woodruff, Chairman, Division of Biology and Agriculture, NRC, June 25, 1929.
5. Cravens, *The Triumph of Evolution*, p. 222.
6. Degler, *In Search of Human Nature*, p. 74.
7. Barkan, *The Retreat of Scientific Racism*, p. 78.
8. Stocking, *Race, Culture, and Evolution*, pp. 142–54.
9. Degler, *In Search of Human Nature*, p. 74.
10. Boas, *Changes in Bodily Form of Descendants of Immigrants*, pp. 2, 7. See also Boas, "The Instability of Human Types," pp. 99–103.
11. Barkan, *The Retreat of Scientific Racism*, p. 77; Degler, *In Search of Human Nature*, pp. 62–3.
12. For an interesting reaction to Boas's study, See the ACS (Rome): Corrado Gini Papers, Mankind Quarterly File, Corrado Gini to Gayre, January 30, 1961: "When Boas first inquiry was published not only Karl Pearson, but many other scholars were very perplexed on the soundness of his results and made reservations and objections, and I was among them. I have then written a paper on this subject published in the Florence 'Archivio per l'Antropologia e l'Etnologia' 1914, vol. XLIV, n. 3–4."
13. Degler, *In Search of Human Nature*, p. 74.
14. Cravens, *The Triumph of Evolution*, pp. 92–5.
15. Spiro, "Nordic vs. anti-Nordic," p. 39.
16. Eckland, "Genetics and Sociology," p. 174.
17. Spiro, "Nordic vs. anti-Nordic," p. 37; Barkan, *The Retreat of Scientific Racism*, p. 95.
18. Benedict, *Patterns of Culture*, p. 12; Degler, *In Search of Human Nature*, p. 206.
19. Buss, *Evolutionary Psychology*, p. 26; Stocking, *Race, Culture, and Evolution*, p. 306.
20. Barkan, *The Retreat of Scientific Racism*, p. 132; Alcock, *The Triumph of Sociobiology*, pp. 131–4.
21. Kroeber, "The Superorganic," p. 169.
22. Kroeber, "Inheritance by Magic," p. 35; Degler, *In Search of Human Nature*, p. 147.
23. Boakes, *From Darwin to Behaviorism*, p. 229. See also Stocking, *Race, Culture, and Evolution*, p. 287 for a somewhat less dramatic statement of essentially the same points.
24. Ward, *Outlines of Sociology*, p. 110.
25. Murdock, "The Science of Culture," p. 200; Buss, *Evolutionary Psychology*, p. 26.
26. Stocking, *Race, Culture, and Evolution*, pp. 287–8.

27. Boas Letter from Madison Grant to Maxwell Perkins, March 7, 1923: Charles Scribner's Sons Archives, Department of Rare Books and Special Collections, Princeton University Libraries, Princeton, N.J., in Spiro, "Nordic vs. anti-Nordic," p. 43.

28. Spiro, "Nordic vs. anti-Nordic," p. 43.

29. Stocking, *Bones, Bodies, Behavior*, p. 183.

30. Barkan, *The Retreat of Scientific Racism*, p. 92.

31. Boakes, *From Darwin to Behaviorism*, pp. 137, 173, 175; Degler, *In Search of Human Nature*, p. 152; Thorpe, *The Origins and Rise of Ethology*, p. 95.

32. Watson, *Behaviorism*, pp. 4–5, 6, 94, 104, 112–13, 136, 164, 238; Degler, *In Search of Human Nature*, p. 155; Boakes, *From Darwin to Behaviorism*, p. 225; Pinker, *The Blank Slate*, p. 19.

33. Watson, *Behaviorism*, revised ed., quoted in Krantz and Allen, "The Rise and Fall of McDougall's Instinct Doctrine," p. 333.

34. Plotkin, *Evolutionary Thought in Psychology*, p. 60.

35. Degler, *In Search of Human Nature*, p. 158.

36. Yerkes Papers, 1.11.22 Correspondence Files, Dr. Zing Yang Kuo. Kuo to Yerkes, n.d.

37. Kuo, "Giving Up Instincts in Psychology," pp. 645–64; Kuo, "The Net Result of the Anti-Heredity Movement in Psychology," p. 197.

38. Buss, *Evolutionary Psychology*, pp. 24–5.

39. Cravens, *The Triumph of Evolution*, p. 215.

40. Krantz and Allen, "The Rise and Fall of McDougall's Instinct Doctrine," p. 327.

41. Huxley, "Ils n'ont que de l'âme: an essay on bird-mind," p. 108.

42. Huxley, "Progress, Biological and Other," p. 51.

43. McDougall, *Introduction to Social Psychology*, 14th ed., p. 456.

44. DUA: William Few Papers, Correspondence April 1–13, 1926. William McDougall to Few, before April 13, 1926.

45. Krantz and Allen, "The Rise and Fall of McDougall's Instinct Doctrine," p. 326.

46. Carmichael, "Some Historical Roots of Present-Day Animal Psychology," 50–1; McCurdy, "William McDougall," p. 113.

47. McDougall, *Is America Safe for Democracy?*, p. 164.

48. McDougall with Watson, *The Battle of Behaviorism*, p. 72.

49. Cravens, *The Triumph of Evolution*, p. 218.

50. DUA: McDougall Papers, Professional Correspondence, July 1931–October 1933. William McDougall, ca. March 1933, Letter to the Editor of the *Fortnightly Review*.

51. DUA: McDougall Papers, Professional Correspondence, January 1935–September 1935. McDougall to George Vaughan, January 29, 1935.

52. DUA: McDougall Papers, Professional Correspondence, January 1936–May 1936. McDougall to Livingston Welch, February 6, 1936. McDougall's adherence to Nordic racial ideology did not win him many fans, either. DUA: William Few Papers, Correspondence June 9–18, 1926. Albert Knudson to Dean Wdmund Soper, June 18, 1926. McDougall "is reported to be a somewhat extreme representative of the Nordic type of thought . . . he has written some articles endorsing that standpoint." McDougall's earlier book *Group Mind*

(New York, London: G.P. Putnam's Sons, 1920) had received "stinging reviews" for its validation of racial stereotypes and hierarchies. Newbold, "William McDougall, MD, FRS," p. 77.

53. BLM, C-B 935: Samuel Jackson Holmes Papers, Carton 1: "Heredity and Marriage," ca. 1940.
54. Holmes, *Studies in Evolution and Eugenics*, p. 116.
55. BLM, C-B 935: Samuel Jackson Holmes Papers, Carton 1: "Heredity and Marriage," ca. 1940.
56. Holmes, "Nature versus Nurture in the Development of the Mind," pp. 245–9.
57. Boakes, *From Darwin to Behaviorism*, p. 199; Elliot, "Robert Mearns Yerkes," pp. 487–9.
58. "The Mental Age of Americans," pp. 213–15; no. 413 (November 1, 1922): 246–8; no. 414 (November 8, 1922): 275–7; no. 415 (November 15, 1922): 297–8; no. 416 (November 22, 1922): 328–30; no. 417 (November 29, 1922): 9–11.
59. Yerkes, *Almost Human*, p. xv.
60. YUA, Yerkes Papers: Box 18, File 313: Henry Pratt Fairchild file. Yerkes to Fairchild, February 27, 1926.
61. YUA, Yerkes Papers: Box 80, File 1518: Yerkes to President Hoover, December 1, 1930.
62. Elliot, "Robert Mearns Yerkes," p. 489. Yerkes's professorship at Yale was designated as "Professor of Comparative Psychobiology and Director of the Yale Laboratories of Primate Biology."
63. YUA, Yerkes Papers: Box 5, File 84: Edwin Boring File. Yerkes to Boring, November 3, 1932.
64. "Meetings," p. 198.
65. Terman's autobiography, in *History of Psychology in Autobiography*, Vol. 2, p. 330.
66. Spiro, "Nordic vs. anti-Nordic," p. 45.
67. Miller, "The Primate Basis of Human Sexual Behavior," p. 380.
68. Miller, "Some Elements of Sexual Behavior in Primates," p. 274.
69. Degler, *In Search of Human Nature*, p. 187.
70. Cravens, *The Triumph of Evolution*, pp. 210–11.

7 Evolutionary Psychology under Attack

1. Degler, *In Search of Human Nature*, pp. 162, 188, 201.
2. Degler, *In Search of Human Nature*, pp. 201–2; Cravens, *The Triumph of Evolution*, p. 226.
3. For a discussion of eugenics and the Left, see Paul, "Eugenics and the Left," pp. 567, 582.
4. Degler, *In Search of Human Nature*, p. 162; Cravens, *The Triumph of Evolution*, p. 192.
5. Dunlap, "Are There any Instincts?" 309.
6. Jennings, *Prometheus; or Biology and the Advancement of Man*; Castle, "Eugenics," pp. 1031–2; Pearl, "The Biology of Superiority," pp. 257–66;

Morgan, *Evolution and Genetics*; Conklin, "Some Recent Criticisms of Eugenics," pp. 61–5; Cravens, *The Triumph of Evolution*, p. 179.

7. APS Archives: Raymond Pearl Papers, File: East, Edward Murray, Box 4–Box 5, Folders #1–9. Pearl to East, November 25, 1927. The letter refers to Pearl, "The Biology of Superiority," pp. 257–66.

8. Tucker, *The Funding of Scientific Racism*, p. 31; Davenport and Steggerda, *Race Crossing in Jamaica*, p. iii.

9. Davenport and Steggerda, *Race Crossing in Jamaica*, pp. 469, 471–2; Tucker, *The Funding of Scientific Racism*, p. 50; Barkan, *The Retreat of Scientific Racism*, p. 345.

10. APS Archives: Herbert Spencer Jennings Papers, File: Davenport, C.B. 40: 1899–1930. Davenport to Jennings, June 17, 1930.

11. NAS: Biology and Agriculture, Committee on Human Heredity, 1930–1939, General, C.G. Campbell to Woodruff, March 30, 1929.

12. NAS: Biology and Agriculture, Committee on Human Heredity, 1930–1939, General, L.L. Woodruff to C.E. Allen, May 24, 1929; C.G. Campbell to L.L. Woodruff, June 25, 1929; C.E. Allen to C.G. Campbell, July 2, 1929.

13. NAS: Biology and Agriculture, Committee on Human Heredity, 1930–1939, General, Meeting of the Committee on Human Heredity. Minutes. February 8, 1930; Report of the second meeting of the Committee on Human Heredity. Davenport to Ludwig Hektoen, March 17, 1930.

14. NAS: Biology and Agriculture, Committee on Human Heredity, 1930–1939, General, Charles Davenport, Chairman, Committee on Human Heredity, February 19, 1932.

15. Barkan, *The Retreat of Scientific Racism*, p. 344.

16. Degler, *In Search of Human Nature*, p. 202.

17. SAA: Ales Hrdlicka Papers, Box 8: American Eugenics Society, 1926–1932, 1940; Letter from Henry F. Perkins, President of the American Eugenics Society, March 7, 1932. In another example, Clarence Campbell wrote to Davenport that he feared the *Eugenical News* might have to cease publication due to the Eugenic Research Association's lack of funds. APS Archives: Charles Benedict Davenport Papers, Series I. File: Campbell, C.G. (Clarence Gorden), 1868–1956, Folders 1–2: Campbell to Davenport, June 28, 1932.

18. Stocking, *Race, Culture, and Evolution*, pp. 296, 301, 305, 306.

19. Stocking, *Race, Culture, and Evolution*, p. 288.

20. See, for instance, McDougall, "Motives in the Light of Recent Discussion," pp. 277–93; McDougall, "The Use and Abuse of Instinct in Social Psychology," pp. 285–333.

21. C.H. Judd, "Evolution and Consciousness," pp. 77–8, 80, 84–5, 88–90; Degler, *In Search of Human Nature*, p. 157.

22. For example, Leonard Carmichael remembered an incident in the mid-1920s when he took his oral doctoral examination at Harvard. Present were McDougall, Walter Dearborn, and G.H. Parker. When Dearborn led Carmichael into the examination, he told him, "Say exactly what seems correct to you. We will agree with you and not with McDougall." Leonard Carmichael's autobiography, in Boring and Lindzey, *A History of Psychology in Autobiography*, Vol. 5, p. 36.

23. Murphy, *Historical Introduction to Modern Psychology*, pp. 336–9.
24. McDougall with Watson, *The Battle of Behaviorism*, p. 26.
25. H.A. Murray's autobiography, in Boring and Lindzey, *A History of Psychology in Autobiography*, Vol. 5, p. 292.
26. A.A. Roback, *A History of American Psychology*, revised ed., p. 168. DUA: McDougall Papers, Professional Correspondence, May 1938–December 1938. Harry Helson to William McDougall, September 15, 1938.
27. Boakes, *From Darwin to Behaviorism*, pp. 210–11.
28. McDougall, "Objection and Reproof," p. 68.
29. Schoen, "Instinct and Man," p. 531.
30. Schoen, "Instinct and Man," pp. 531, 533, 534.
31. R.L. Duffus, "Social Behavior among Monkeys," p. BR4.
32. Kuhn Jr., "Forum on Behavior Most Misbehaved," p. 10.
33. Jones, "Psychology, History, and the Press," pp. 931–40.
34. Raglan, "Incest and Exogamy," pp. 168, 169, 178.
35. Murdock, "The Science of Culture," pp. 200, 201, 202, 207, 208, 210–11, 213–14.
36. V.F. Calverton, "The Compulsive Basis of Social Thought," pp. 696, 697, 698, 703.
37. For an example of this view, Tisdale refers to F.H. Hankins, *An Introduction to the Study of Society*, p. 260.
38. Tisdale, "Biology in Sociology," pp. 34, 39. Tisdale cites Ross's *Principles of Sociology*, p. 504 as an example of his claim that avoidance of incest is a human instinct.
39. See also Huxley, *Evolution*.
40. Huxley, "The Uniqueness of Man," pp. 475, 478, 489, 490, 493, 494.
41. For more on this, see Cravens, *The Triumph of Evolution*.
42. Thorndike, through his research on intelligence in twins, found that "the mental likeness found in the case of twins . . . are due, to at least nine tenths of their amount, to original nature. They justify the emphasis put upon the magnitude of heredity as a cause of the mental differences amongst men . . ." *Measurement of Twins*, p. vii.
43. Witty and Lehman, "The Dogma and Biology of Human Inheritance," pp. 552, 563.
44. NARA, Record no. 592.7c 17/47.
45. Hogben, *Genetic Principles in Medicine and Social Science*, p. 94.
46. Hogben, *Genetic Principles in Medicine and Social Science*, p. 93.
47. Kaempffert, "Review of Hogben, *Genetic Principles in Medicine and Social Science*," p. BR17.
48. Kaempffert, "New Horizons for Science," p. BR1. The reader will note that the coeditor of *Science for a New World* was the same J. Arthur Thompson who Gerrit Miller Jr. criticized several years before.
49. APS Archives, BD27: Charles Benedict Davenport Papers, Series I, File: Kaempffert, Waldemar. Kaempffert to Davenport, December 28, 1915; Kaempffert to Davenport January 10, 1916.
50. "Behavior Held Vital to Greater Stature," p. 23.
51. "Says Physical Facts Can Explain the Mind," p. 17.

52. "New Tests Attack Theory of Fixed IQ," p. 32.
53. DUA: William McDougall Papers, Professional Correspondence, October 1934–November 15, 1934. Fort to McDougall, October 15, 1934.

8   The Death of Evolutionary Psychology

1. LOC: Merriam Papers, Box 77, Madison Grant File. Grant to Merriam, November 27, 1933.
2. Bachman, "Theodore Lothrop Stoddard," p. 223; W.K. Gregory to Madison Grant, October 20, 1930: CBD: "William K. Gregory" Folder; T.W. Todd to W.K. Gregory, March 21, 1932: William King Gregory Papers, American Museum of Natural History, New York, "Third International Commission of Eugenics, 1922–1934," Folder, Box 2, in Spiro, "Nordic vs. anti-Nordic," p. 48.
3. APS Archives, BD27: Charles Benedict Davenport Papers, Series I, File: Holmes, Samuel Jackson. Holmes to Davenport, January 8, 1929; Davenport to Holmes, January 14, 1929.
4. APS Archives, BD27: Charles Benedict Davenport Papers, File: Holmes Samuel Jackson. Davenport to Holmes, January 14, 1929.
5. CIW: John Campbell Merriam Papers, Memorandum Files. Memorandum for conversation with Dr. Wissler and Dr. Kidder, December 31, 1928; Cravens, *The Triumph of Evolution*, p. 179.
6. APS Archives, BD27: Charles Benedict Davenport Papers, File: Holmes Samuel Jackson. Davenport to Holmes, January 14, 1929.
7. Cravens, *The Triumph of Evolution*, pp. 179–80.
8. APS Archives: Frederick Osborn Papers, File: Charles Davenport. Osborn to Davenport December 10, 1931.
9. See Muller, "The Dominance of Economics over Eugenics," pp. 138–44.
10. Glass, preface to Muller, *Man's Future Birthright*, p. xi.
11. Reference to Gini, "Response to the Presidential Address," pp. 25–8.
12. "Urge Birth Control to Aid World Peace," p. 21.
13. APS Archives, BD27: Charles Benedict Davenport Papers, Series I, File: Johnson, Roswell Hill. Johnson to Davenport December 2, 1921; Davenport to Johnson, December 5, 1921.
14. APS Archives, BD27: Charles Benedict Davenport Papers, File: Johnson, Roswell Hill. M.J. Exner to Davenport, September 13, 1922. Samuel Holmes, William McDougall, and a number of other eugenicists in this work were also involved in the birth control movement.
15. "Eugenics Committee Offers a Program," p. 12.
16. APS Archives, BD27: Charles Benedict Davenport Papers, File: Johnson, Roswell Hill. Davenport to Johnson, June 18, 1926; Davenport to Johnson, April 29, 1931.
17. AHC: Paul Popenoe Papers, Box 114, Folder 3, 1933–1936: Paul Popenoe to Mrs. Comstock, June 25, 1936.
18. "Birth Control Ban Fought by Doctors," p. 21.
19. UPASC: Collection 2/10 McCormick 1904–1920, FF31. Johnson to S.B. McCormick, June 4, 1918.

20. E.B. Reuter, "Review of *Applied Eugenics*," p. 549.
21. Marianne Kasica, University Archives, University of Pittsburgh. Personal communication. May 21, 2004; University of Pittsburgh, Archives Service Center. Collection 2/10 McCormick 1904–1920, FF31. Johnson to S.B. McCormick, January 4, 1916; Johnson to McCormick, December 15, 1916; S.B. McCormick to Johnson, January 15, 1917; John to McCormick, January 15, 1917.
22. APS Archives, BD27: Charles B. Davenport Papers, Series I, File: Johnson, Roswell H. Johnson to Davenport, March 31, 1932. Johnson to Davenport June 3, 1935; Johnson to Davenport March 31, 1936; "In Memoriam: Roswell H. Johnson," 56.
23. AHC: Paul Popenoe Papers, Box 144, Folder 2: Philip S. Platt to Paul Popenoe, April 25, 1936.
24. AHC: Paul Popenoe Papers, Box 114, Folder 3, 1933–1936: Paul Popenoe to Mrs. Comstock, June 25, 1936.
25. Marquis, *Who's Who in America 1950–1951*, Vol. 26, p. 1405.
26. For growing opposition to the Institute's eugenics lectures, see for example the AHC: Paul Popenoe Papers, Box 144, Folder 2: Bertha Stevens to L.D. Osborn, December 17, 1935.
27. APS Archives, American Eugenics Society Records. Correspondence and Records: Johnson, Roswell Hill. Misc. materials: 14. Prof. Roswell Hill Johnson. Stefan Kühl's observation that Johnson was a "reformist" eugenicist in the mold of Julian Huxley and Frederick Osborn is quite obviously inaccurate. Kühl, *Die Internationale der Rassisten*, p. 109.
28. For more on the American–German eugenic connection, see the excellent work by Kühl, *The Nazi Connection: Eugenics*.
29. APS: Leslie C. Dunn Papers, Paul Popenoe to Leslie Dunn, January 22, 1934; Ludmerer, *Genetics and American Society*, p. 118; Miller, *Terminating the "Socially Inadequate,"* p. 149.
30. Miller, Terminating the "Socially Inadequate," p. 29.
31. AHC: Paul Popenoe Papers, Box 11, Folder 7: Paul Popenoe to Betty Stankowitch Popenoe, July 13, 1931.
32. LOC: John C. Merriam Papers, Box 138: Henry Fairfield Osborn File. H.F. Osborn to Merriam, October 11, 1933. LOC: John C. Merriam Papers, Box 138: Henry Osborn File. Merriam to Henry Osborn, October 29, 1935.
33. APS Archives: Frederick Osborn Papers, File: "Concerning Eugenics." Osborn to Richard Johnson, n.d.
34. CIW Archives: Frederick Osborn to John C. Merriam, October 23, 1934.
35. TSL: Laughlin Papers, "Confidential Note for Dr. Ramos."
36. APS Archives, B/B61: Franz Boas Papers, Professional Correspondence, File: Osborn, Frederick. Osborn to Boas, October 11, 1937.
37. LOC: John C. Merriam Papers, Box 138: Frederick Osborn File. Osborn to Merriam, September 9, 1935; Miller, *Terminating the "Socially Inadequate,"* p. 149.
38. APS Archives: Frederick Osborn Papers, File: Galton Society, Box 7. William K. Gregory to Clarence Campbell, May 6, 1935.
39. Barkan, *The Retreat of Scientific Racism*, p. 275.

40. NAS: Report of the Committee on Human Heredity, March 19, 1935; Report of the Committee on Human Heredity, March 16, 1936.
41. APS Archives: Frederick Osborn Papers, File: Charles Davenport. Davenport to Osborn, September 11, 1935.
42. "Democracy and Eugenics," p. 66.
43. LOC: John C. Merriam Papers, Box 138: Frederick Osborn File. Osborn to Merriam, "Memorandum Re Eugenics Record Office" May 10, 1937; Merriam to Osborn, August 27, 1935; Osborn to Merriam, April 9, 1937.
44. Cravens, *The Triumph of Evolution*, pp. 179–80.
45. APS Archives, B/B61: Franz Boas Papers, Professional Correspondence, File: Osborn, Frederick. Osborn to Boas, June 7, 1938.
46. APS Archives: Frederick Osborn Papers. File: American Eugenics Society 1940, 1976. Pres., Assoc. for Research in Human Heredity to Klineberg, Shapiro, Robinson, Caughey, Snyder, Schweitzer, Fowler, Draper, Berens, December 13, 1946.
47. APS Archives: Frederick Osborn Papers no. 24, Assoc. for Research in Human Heredity. Secretary's book 1938–1940, Minutes of the meeting of the ERA, December 1, 1938. See also the AHC: Paul Popenoe Papers. Maurice Bigelow to Popenoe, September 14, 1949; January 18, 1951; May 2, 1951; January 12, 1952; January 5, 1953. In the January 12, 1952, letter to Popenoe, Bigelow wrote: "A number of important AES members of years gone by have written, *confidentially*, that they dropped out of AES because they could not see AES as worth while. Some wrote that it "never accomplished anything of importance." I guess they are about right so far as eugenics has affected population, either local or national; but I have known so many hundreds of cases where the eugenics idea has worked in families that I believe that it worth while to keep the idea alive. That will be difficult because F.O. [Frederick Osborn] and his leading friends are all population minded."
48. SAA: Ales Hrdlicka Papers; Rudolf Bertheau, Sec., AES to Ales Hrdlicka, January 26, 1940; APS Archives: Frederick Osborn Papers, File: "Concerning Eugenics." Osborn to Albert Blakeslee, April 23, 1940.
49. Osborn, "The American Concept of Eugenics," p. 2.
50. "Plan for Improving Population," p. 63.
51. SAA: Ales Hrdlicka Papers; Ales Hrdlicka to Rudolf Bertheau, January 25, 1940.
52. YUA: Robert Yerkes Papers, Box 80, File 1519: American Eugenics Society, 1930–1952. M.A. Bigelow to Yerkes, May 2, 1950.
53. DUA: William McDougall Papers, Professional Correspondence. May 1938–December 1938. G.H. Estabrooks to William McDougall, September 6, 1938.
54. Dunlap, "Antidotes for Superstitions Concerning Human Heredity," p. 221.
55. SAA: Ales Hrdlicka Papers. Ales Hrdlicka to Rudolf Bertheau, January 31, 1940.
56. SAA: Coon Papers, Carleton Coon to Carleton Putnam, June 17, 1960.
57. Cravens, *The Triumph of Evolution*, p. 241.
58. Yerkes, "Genetic Aspects of Grooming," p. 22.
59. Yerkes, "Primate Cooperation and Intelligence," pp. 264, 266.
60. YUA: Yerkes Papers, Box 80, File 1519: Eugenics. Robert Yerkes, "Statement for Use of the American Eugenics Society," December 22, 1939.

61. YUA: Yerkes Papers, Box 80, File 1519: Eugenics. Ellsworth Huntington to Yerkes, May 7, 1941.

62. Haslerud, preface to Yerkes, *The Mental Life of Monkeys and Apes*, p. vii.

63. Cravens, *Triumph of Evolution*, p. 220.

64. Krantz and Allen, "The Rise and Fall of McDougall's Instinct Doctrine," p. 333. Vitalism referred to the idea that life was self-conscious. Today only humans, and perhaps the highest animals, are thought to possess any meaningful form of self-consciousness.

65. Lashley, "Experimental Analysis of Instinctive Behavior," pp. 443–71; see Thorpe, *The Origins and Rise of Ethology*, p. 48.

66. Thorpe, *The Origins and Rise of Ethology*, p. 49.

67. Westermarck, "Methods in Social Anthropology," p. 227.

68. "Review of S.J. Holmes, *The Eugenic Predicament*," p. 51.

69. Bancroft Library Manuscripts, C-B 935: Samuel Jackson Holmes Papers, Carton 2. For another critical review of this book, see "Review of *The Eugenic Predicament*," pp. 225–8.

70. BLM, BANC MSS C-B 935: Holmes, Samuel J., Box 1, Correspondence A–H. File: Darwin, Leonard. Leonard Darwin to S.J. Holmes, March 14, 1937.

71. Holmes, "The Opposition to Eugenics," 355–7. Holmes made similar arguments five years before, in his book *The Eugenic Predicament*, p. 122ff.

72. BLM, BANC MSS C-B 935: Holmes, Samuel J., Box 1, Correspondence A–H. File: C.M. Goethe. C.M. Goethe to S.J. Holmes, August 5, 1960; C.M. Goethe to S.J. Holmes January 7, 1963.

73. BLM, BANC MSS C-B 935: Holmes, Samuel J., Box 1, Correspondence A–H. File: C.M. Goethe. C.M. Goethe to Ann F. Isaacs, April 14, 1964.

74. Duncan and Duncan, "Shifts in Interests of American Sociologists," p. 210, table 1.

75. Ogburn, "Trends in Social Science," p. 259.

76. Krantz and Allen, "The Rise and Fall of McDougall's Instinct Doctrine," p. 335.

77. McCurdy, "William McDougall," p. 123. The study referred to was Schoolland et al., *Are There any Innate Behavior Tendencies?* pp. 25, 219–87.

78. Herrnstein, "Nature as Nurture," pp. 24–5; Degler, *In Search of Human Nature*, pp. 165, 205; Jones, "Psychology, History, and the Press," pp. 931–40; Krantz and Allen, "The Rise and Fall of McDougall's Instinct Doctrine," 326–38.

79. NAS Archives: Division of Biology and Agriculture, Committee on Human Heredity, 1930–1939. General, February 1938, revised in April 1938.

80. NAS Archives: Division of Biology and Agriculture, Committee on Human Heredity, 1930–1939. General, Emund W. Sinnott to R.G. Harrison, November 2, 1939.

81. NAS Archives: Division of Biology and Agriculture, Committee on Human Heredity, 1930–1939. General, R.E. Coker to Ross Harrison, December 21, 1939.

82. Kevles and Hood, *The Code of Codes*.

83. Dunn had been a member of the Galton Society. APS Archives: Frederick Osborn Papers, File: Charles Davenport. Osborn to Davenport, May 6, 1931.

84. Dunn, "Cross Currents in the History of Human Genetics," pp. 1, 3.

85. Pauly, *Biologists and the Promise of American Life*, p. 226.
86. Report of the Committee on Human Heredity, April 1, 1943; Report of the Committee on Human Heredity, February 1944. We should note that the Committee also funded several studies on twins conducted by researchers at the University of Chicago and at Harvard, attempting to assess the relative contributions of environment and heredity to intelligence.
87. Bowler, *Evolution*, p. 292; Miller, "A Review of *Sexual Selection and Human Evolution*."
88. Kandel, "A New Intellectual Framework for Psychiatry," pp. 457–69.

9 Lost in the Wilderness

1. A good example of the difficulties that anyone advocating evolutionary psychology encountered in publishing is found in a letter from F. Ivan Nye, the editor of *Marriage and Family Living*, to Paul Popenoe, concerning an article he attempted to get published: "The evaluations of two readers are now in. They do not favor publication of the present paper because they feel current evidence is generally against the position that many of these characteristics (for example, introversion-extroversion) are hereditary." AHC: Paul Popenoe Papers, Box 122, Folder 9, "Heredity Counseling": F. Ivan Nye to Popenoe, August 3, 1960.
2. APS Archives, BD27: Charles Benedict Davenport Papers, Series I, File: Castle, William E. (William Ernest), 1867–1962, Folder 7: "Report of Progress under Grant no. 562," September 1, 1910.
3. Phineas Whiting was married to Anna Rachel Whiting, who worked with Josephine Naly on sexual selection studies at the University of Pittsburgh.
4. CIW: John Campbell Merriam Correspondence, 1922: Davenport to Merriam, July 13, 1922; Davenport to Merriam, August 15, 1922; CIW: John Campbell Merriam Papers, 1920–1930, Box 3: Conference with Dr. MacDowell in New York, October 19–20, 1922.
5. National Academy of Sciences, "Clarence C. Little," p. 248.
6. The *Michigan Alumnus*, commenting on Little's resignation as president of the University of Michigan, said that "As a biologist, Dr. Little has been a strong advocate of race betterment programs and the science of eugenics and his courageous statements of his views on controversial subjects . . . have led to criticism in some quarters." Holstein and Dupuy, *The First Fifty Years at the Jackson Laboratory*, p. 11; Rader, "C.C. Little and the Jackson Laboratory Archives."
7. Laurence, "Sees a Super-Race Evolved by Science," p. 40.
8. Holstein and Dupuy, *The First Fifty Years at the Jackson Laboratory*, p. 21.
9. YUA: Yerkes Papers, Box 2, File 22: W.C. Allee Files. Allee to Yerkes, June 15, 1945.
10. YUA: Yerkes Papers, Box 70, File 1334: Bar Harbor Conference; Holstein and Dupuy, *The First Fifty Years at the Jackson Laboratory*, p. 187; Degler, *In Search of Human Nature*, p. 218.
11. Gini, "J. P. Scott, the Social Behavior of Dogs and Wolves," p. 272.

12. Holstein and Dupuy, *The First Fifty Years at the Jackson Laboratory*, p. 197. Interest in the unusual intelligence of dogs has since revived.
13. Eaton, "An Historical Look at Ethology," p. 171; Krantz and Allen, "The Rise and Fall of McDougall's Instinct Doctrine," 332.
14. Yerkes, "The Scope of Science," pp. 461–3. See also "The Method and Scope of Science," p. 126.
15. YUA: Yerkes Papers, Box 43, Folder 925: Niko Tinbergen File. Tinbergen to Yerkes, April 30, 1938.
16. YUA: Yerkes Papers, Box 43, Folder 925: Niko Tinbergen File. Tinbergen to Yerkes, October 24, 1938.
17. YUA: Yerkes Papers, Box 43, Folder 925: Niko Tinbergen File. Tinbergen to Yerkes, October 18, 1946.
18. YUA: Yerkes Papers, Box 20, File 356: Karl von Frisch. Yerkes to von Frisch, January 9, 1951.
19. Garcia, "The Psychological Literature in Konrad Lorenz's Work," pp. 2, 3, 5, 6, 15.
20. Gini, "J. P. Scott, the Social Behavior of Dogs and Wolves," p. 272.
21. Gini, *Corso di Sociologia*, p. 556.
22. Barkan, *The Retreat of Scientific Racism*, p. 342.
23. Montagu, "The New Litany of 'Innate Depravity,'" p. 11; Degler, *In Search of Human Nature*, p. 209.
24. Carmichael, "Some Historical Roots of Present-Day Animal Psychology," pp. 65, 182.
25. SAA: Coon Papers, Coon to Putnam, June 17, 1960.

Conclusion

1. Spiro, "Nordic vs. anti-Nordic," p. 40.
2. Jones, "Psychology, History, and the Press," pp. 931–40; Cohen, *Psychologists on Psychology*, p. 322.
3. Garcia, "The Psychological Literature in Konrad Lorenz's Work," p. 3.
4. Lorenz, *The Foundations of Ethology*, pp. 68–71.
5. Cartwright, *Evolution and Human Behavior*, p. 324. Oddly enough, Niko Tinbergen would himself have to deal with accusations of Nazi sympathies. In the 1960s, while on a lecture tour of Canada, Tinbergen's steps were dogged by a group of Canadian Maoist students who accused him of Nazism because he published an article in a German periodical, *Zeitschrift für Tierpsychologie*, during World War II. At one point the students even succeeded in breaking up one of his lectures. They apparently did not know that Tinbergen had resigned from the University of Leyden during the war in protest against the dismissal of his Jewish colleagues there by the Germans. Tinbergen spent two years in a German concentration camp as a result. Cohen, *Psychologists on Psychology*, pp. 321, 327.
6. Hess, "Comparative Psychology," pp. 239, 242, 751.
7. Cohen, *Psychologists on Psychology*, p. 317.

8. Thorpe, *The Origins and Rise of Ethology*, pp. 49–51, 88; YUA: Yerkes Papers, Box 43, Folder 925: Niko Tinbergen File. Tinbergen to Yerkes, October 18, 1946. William James was lucky enough to die before the anti-instinct onslaught commenced. Thus, he could only be condemned *in absentia*, as having caused "unparalleled harm to psychology" (Henry A. Murray's autobiography, in Boring and Lindzey, *A History of Psychology in Autobiography*, p. 293). As we have seen, William McDougall, as well as others, was not so fortunate.

9. McCurdy, "William McDougall," p. 123.

10. Krantz and Allen, "The Rise and Fall of McDougall's Instinct Doctrine," p. 332. See Beach, "The Descent of Instinct," pp. 401–10.

11. Blough and Millward, "The Comparative Psychology of Learning," p. 108.

12. Carmichael, "Some Historical Roots of Present-Day Animal Psychology," p. 187.

13. Hirsch, "Behavior-Genetics, or 'Experimental' Analysis," p. 120.

14. Morris, "The Rigidification of Behaviour," p. 327.

15. Degler, *In Search of Human Nature*, p. 224.

16. Eckland, "Genetics and Sociology," p. 173.

17. Eckland, "Genetics and Sociology," p. 173.

18. Eckland, "Genetics and Sociology," p. 179.

19. Eckland, "Genetics and Sociology," p. 176.

20. McCurdy, "William McDougall," p. 123.

21. Eaton, "An Historical Look at Ethology," p. 176.

22. Eaton, "An Historical Look at Ethology," p. 177.

23. Eaton, "An Historical Look at Ethology," p. 179.

24. Haldane, *The Causes of Evolution*, p. 170.

Bibliography

Archival Sources

ACS. Corrado Gini Papers.
AHC. Paul Popenoe Papers.
APS. Frederick Osborn Papers.
APS. Charles Benedict Davenport Papers.
APS. Herbert Spencer Jennings Papers.
APS. Raymond Pearl Papers.
BLM. C-B 935. Samuel Jackson Holmes Papers.
CIW. John Campbell Merriam Papers.
CIW. Robert Yerkes Files.
CIW. Roswell H. Johnson File, 1902, 1904–1906, 1919.
DUA. William Few Papers.
LOC. John Campbell Merriam Papers.
NARA. Record no. 592.7c 17/47. International Conferences (Congresses), 1918–1939.
NAS. Division of Biology and Agriculture. Committee on Human Heredity, 1930–1939.
NAS. Report of the Committee on Human Heredity.
SAA. Ales Hrdlicka Papers.
SAA. Carleton Coon Papers.
SI. Gerrit Miller Papers.
TSL. Harry Laughlin Papers.
UPASC. Collection 2/10 S.B. McCormick 1904–1920, FF31.
YUA. Robert M. Yerkes Papers.

Primary Sources

"An Ideal Wife's Qualities." *Physical Culture* 34, 4 (October 1915): 60.
"Behavior Held Vital to Greater Stature." *New York Times* (July 10, 1934): 23.
"Birth Control Ban Fought by Doctors." *New York Times* (February 14, 1931): 21.
"Catechism of the Catholic Church," no. 867. http://www.vatican.va/archive/catechism/p123a9p3.htm. Accessed September 23, 2006.
"Democracy and Eugenics." *New York Times* (May 16, 1937): 66.
"In Memoriam: Roswell H. Johnson." *Eugenics Quarterly* 15, 1 (March 1968): 56.

"Meetings." *American Anthropologist* (New Series) 32, 1 (January–March 1930): 189–203.

"New Tests Attack Theory of Fixed IQ." *New York Times* (July 17, 1938): 32.

"Plan for Improving Population: Outlined at the Seventh International Congress of Geneticists." *Eugenics Review* 28, 3 (March 1939): 63.

"Says Physical Facts Can Explain the Mind." *New York Times* (September 5, 1936): 17.

"The Ideal Husband." *Physical Culture* 35, 3 (March 1916): 38–43.

"Urge Birth Control to Aid World Peace." *New York Times* (November 12, 1921): 21.

Alighieri, Dante. Trans. Marvin Richardson Vincent. *The Divine Comedy of Dante: The Inferno.* New York: Charles Scribner's Sons, 1904.

Allee, Warder Clyde. *Animal Aggregations: A Study in General Sociology.* Chicago: University of Chicago Press, 1931.

Allen, E., B. Beckwith, J. Beckwith, S. Chorover, D. Culver, M. Duncan, J.S. Gould, J.R. Hubbard, H. Inouye, A. Leeds, R. Lewontin, C. Madansky, L. Miller, R. Pyeritz, M. Rosenthal, and H. Schreier. "Against Sociobiology." *New York Review of Books* 22 (1975).

Allen, Garland. "Is a New Eugenics Afoot?" *Science* 294 (2001): 59–61.

Alverdes, Friedrich. *Social Life in the Animal World.* New York: Kraus Reprint, 1969 [1927].

———. *The Psychology of Animals in Relation to Human Psychology.* London: Routledge, 1999 [1932].

Ashley-Montagu, M.F. "Selfish Scientists." *New York Times* (January 28, 1940): E9.

Beach, Frank A. "The Descent of Instinct." *Psychological Review* 62 (1955): 401–10.

Benedict, Ruth. *Patterns of Culture.* Boston: Houghton Mifflin, 1958.

Bingham, Harold. *Sex Development in Apes.* Baltimore, Md.: Johns Hopkins Press, 1928.

Blough, Donald S. and Richard B. Millward. "The Comparative Psychology of Learning." In Paul R. Farnsworth and Olga McNemar, eds. *Annual Review of Psychology.* Palo Alto, Calif.: Annual Reviews, 1965.

Boas, Franz. *Changes in Bodily Form of Descendants of Immigrants.* Washington, D.C.: Government Printing Office, 1911.

———. "The Instability of Human Types." In Gustav Spiller, ed. *Papers on Interracial Problems Communicated to the First Universal Races Congress Held at the University of London, July 26–29, 1911.* Boston: Ginn and Co., 1912.

———. "Eugenics." *The Scientific Monthly* 3, 5 (November 1916): 471–8.

Brigham, Carl C. Foreword by Robert M. Yerkes. *A Study of American Intelligence.* Princeton: Princeton University Press, 1923.

Buss, David M. "Sex Differences in Human Mate Preferences: Evolutionary Hypotheses Tested in 37 Cultures." *Behavioral and Brain Sciences* 12 (1989): 1–49.

Buss, David M. and Michael Barnes. "Preferences in Human Mate Selection." *Journal of Personality and Social Psychology* 50, 3 (1986): 559–70.

Buss, David M., Randy J. Larsen, Drew Westen, and Jennifer Semmelroth. "Sex Differences in Jealousy: Evolution, Physiology, and Psychology." *Psychological Science* 3, 4 (July 1992): 251–5.

Calverton, V.F. "The Compulsive Basis of Social Thought: As Illustrated by the Varying Doctrines as to the Origins of Marriage and the Family." *The American Journal of Sociology* 36, 5 (March 1931): 689–734.

Carmichael, Leonard. Autobiography. In E.G. Boring and G. Lindzey, eds. *A History of Psychology in Autobiography*. Vol. 5. New York: Appleton-Century-Crofts, 1967.

———. "Some Historical Roots of Present-Day Animal Psychology." In Benjamin B. Wolman, ed. *Historical Roots of Contemporary Psychology*. New York: Harper and Row, 1968.

Carr-Saunders, A.M. *Eugenics*. New York: Henry Holt and Co.; London: Williams and Norgate, Ltd., 1926.

Case, Clarence Marsh. "Instinctive and Cultural Factors in Group Conflicts." *The American Journal of Sociology* 28, 1 (July 1922): 1–20.

Castle, William E. "Eugenics," *Encyclopedia Britannica*. 13th ed. London/New York: The Encyclopedia Britannica Co./The Encyclopedia Britannica, 1926, pp. 1031–2.

Cattell, Jacques, ed. *American Men of Science: A Biographical Directory*. 8th ed. Lancaster, Penn.: Science Press, 1949.

Chorover, S.L. *From Genesis to Genocide: The Meaning of Human Nature and the Power of Behavioral Control*. Cambridge, Mass.: MIT Press, 1979.

Cohen, David. *Psychologists on Psychology*. New York: Taplinger Publishing Company, 1977.

Conklin, Edwin G. "Some Recent Criticisms of Eugenics." *Eugenical News* 13 (1928): 61–5.

Craig, Wallace. "Appetites and Aversions as Constituents of Instincts." *The Biological Bulletin* 34, 2 (1918): 91–107.

———. "Why Do Animals Fight?" *International Journal of Ethics* 31, 3 (April 1921): 264–78.

Crawford, M.P. "Cooperative Behavior in Chimpanzee." Read at the Ann Arbor meeting of the American Psychological Association. September 5, 1935.

Darwin, Charles. *Expression of the Emotions in Man and the Animals*. New York: Philosophical Library, 1955 [1872].

———. *The Descent of Man*. 2nd ed. Amherst, N.Y.: Prometheus Books, 1998 [1874].

Darwin, Leonard. "Mate Selection." *The Eugenics Review* 15 (1923): 459–71.

Davenport, C.B. and Morris Steggerda. *Race Crossing in Jamaica*. Washington, D.C.: Carnegie Institution of Washington, 1929.

Davey, John. *Account of the Interior of Ceylon, and of Its inhabitants: With Travels in that Island*. London: Longman, Hurst, Rees, Orme, and Brown, 1821.

Drew, Frank. "Notes." *Pedagogical Seminary* (1907): 504–5.

Duffus, R.L. "Social Behavior among Monkeys." *New York Times* (July 17, 1932): BR4.

Duncan, H.G. and Winnie Leach Duncan. "Shifts in Interests of American Sociologists." *Social Forces* 12, 2 (December 1933): 209–12.

Dunlap, Knight. "The Significance of Beauty." *Psychological Review* 25 (1918): 191–213.

———. "Are There any Instincts?" *Journal of Abnormal Psychology* 14 (1919–1920): 307–11.

Dunlap, Knight. "The Social Need for Scientific Psychology." *The Scientific Monthly* 11, 6 (December 1920): 502–17.

———. *Personal Beauty and Racial Betterment*. St. Louis: C.V. Mosby Co., 1920.

———. "Antidotes for Superstitions Concerning Human Heredity." *The Scientific Monthly* 51, 3 (September 1940): 221–5.

Dunn, Leslie C. "Cross Currents in the History of Human Genetics." *American Journal of Human Genetics* 14 (1963): 1–13.

East, Edward M. *Heredity and Human Affairs*. New York and London: Charles Scribner's Sons, 1927.

Eaton, R.L. "An Historical Look at Ethology: A Shot in the Arm for Comparative Psychology." *Journal of the History of the Behavioral Sciences* 6 (April 1970): 176–87.

Eckland, Bruce. "Genetics and Sociology: A Reconsideration." *American Sociological Review* 32, 3 (April 1967): 173–94.

Ellis, Havelock. *Studies in the Psychology of Sex: Sexual Selection in Man*. Philadelphia: F.A. Davis, 1925 [1905].

———. *Studies in the Psychology of Sex: Sex in Relation to Society*. Vol. IV, 3rd ed. New York: Random House, 1936.

Ellwood, Charles A. *Sociology in Its Psychological Aspects*. 2nd ed. New York and London: D. Appleton and Co., 1919.

———. "Obituary: William McDougall: 1871–1938." *The American Journal of Sociology* 44, 4 (January 1939): 555.

Emlen, S.T., P.H. Wregem, and N.J. Demong. "Making Decisions in the Family: an Evolutionary Perspective." *The American Scientist* 83 (March/April 1995): 148–57. Reprinted in P.W. Sherman and J. Alcock, eds. *Exploring Animal Behavior: Readings from* American Scientist. 2nd ed. Sunderland, Mass.: Sinauer Associates, 1997.

Féré, Charles. *L'instinct sexuel: Évolution et dissolution*. Paris: Félix Alcan, 1899.

Ferrero, Guigliemo. "The Problem of Women from a Biosocial Point of View." *Monist* 4 (1894): 261–74.

Fisher, Ronald A. *The Genetical Theory of Natural Selection*. Oxford: Clarendon Press, 1930.

Fol, Hermann. "La resemblance entre epoux." *Revue Scientifique* 47 (1891): 47–9.

Gallichan, Walter. *The Psychology of Marriage*. New York: Frederick A. Stokes Company, 1918.

Gallup, Jr., Gordon G., and Rebecca L. Burch. "Semen Displacement as a Sperm Competition Strategy in Humans." *Evolutionary Psychology* 2, 8 (February 2004): 12–23.

Galton, Francis. "Composite Portraits, Made by Combining Those of Many Different Persons into a Single Resultant Figure." *The Journal of the Anthropological Institute of Great Britain and Ireland* 8. (June 24, 1879): 132–44.

———. *Inquiries into Human Faculty and its Development*. New York: AMS edition, 1973 [1907].

Gangestead, S.W., R. Thornhill, and R.A. Yeo. "Facial Attractiveness, Developmental Stability, and Fluctuating Asymmetry." *Ethology and Sociobiology* 15 (1994): 73–85.

Geddes, Patrick and J.Arthur Thompson. *The Evolution of Sex*. Revised ed. London: The Walter Scott Publishing Co., Ltd., 1911.

Gesell, Arnold L. "Jealousy." *American Journal of Psychology* 17, 4 (October 1906): 437–96.

Gini, Corrado. "Response to the Presidential Address." *Proceedings of the 3rd International Congress of Eugenics*. Baltimore: The Williams & Wilkins Company, 1934, 25–8.

———. "Notes: Beauty, Marriage and Fertility." *Human Biology* (1938): 576.

———. "J. P. Scott, the Social Behavior of Dogs and Wolves: An Illustration of Socio-Biological Systematics." *Genus* 12, 1–4 (1956): 272.

———. *Corso di Sociologia*. Rome: Edizioni "Ricerche," 1957.

Glass, Bentley. Preface to Hermann J. Muller. *Man's Future Birthright: Essays on Science and Humanity*. Ed. Elof Axel Carlson. Albany, N.Y.: State University of New York Press, 1973.

Glass, Jay D. *Soldiers of God: Primal Emotions and Religious Terrorists*. Corona del Mar, California: Donnington Press, 2003.

Goetz, A. T., and Shackelford, T. K "Sexual Coercion and Forced In-Pair Copulation as Sperm Competition Tactics in Humans." *Human Nature* 17, 3 (Fall 2006): 265–82.

Gould, Stephen Jay. "Biological Potential vs. Biological Determinism." *Natural History* 85 (1976): 12+.

———. "The Diet of Worms and the Defenestration of Prague." *Natural History* 105 (1996): 18–24; 64–7.

Grammer, K. and R. Thornhill. "Human Facial Attractiveness and Sexual Selection: The Roles of Averageness and Symmetry." *Journal of Comparative Psychology* 108 (1994): 233–42.

Grillenzoni, Carlo A. "I caratteri del fisico e del vestire come fattori demografici." *Metron* 10, 3 (1932): 131–84.

Groos, Karl. *Play of Animals*. New York: D. Appleton and Co., 1898.

Haldane, J.B.S. *The Causes of Evolution*. Ithaca, N.Y.: Cornell University Press, 1966 [1932].

Hall, G. Stanely. "Mental Science." *Science* (New Series) 20, 511 (October 14, 1904): 481–90.

———. "Weighs Heart and Emotion to Tell the Scientific Secret of Love." *Boston American* (June 20, 1907).

———. *Adolescence: Its Psychology and Its Relation to Physiology, Anthropology, Sociology, Sex, Crime, Religion and Education*. Vol. II. New York: D. Appleton and Company, 1907.

———. "Eugenics: Its Ideals and What It Is Going to Do." *Religious Education* (June 1911): 152–9.

Hamilton, W.D. "The Genetical Evolution of Social Behaviour. I." *Journal of Theoretical Biology* 7 (1964): 1–16.

———. "The Genetical Evolution of Social Behaviour. II." *Journal of Theoretical Biology* 7 (1964): 17–52.

Hankins, Frank H. *An Introduction to the Study of Society*. New York: The Macmillan Company, 1935.

Harlow, Harry F. "The Nature of Love." *American Psychologist* 13, (1958): 573–685. http://psychclassics.yorku.ca/Harlow/love.htm. Accessed November 3, 2006.

Haslerud, George M. Preface to Robert Yerkes. *The Mental Life of Monkeys and Apes.* Demar, N.Y.: Scholars; Facsimiles and Reprints, 1979 [1916].

Hastings, Milo. "Dress and Its Relation to Sex." *Physical Culture* 32, 2 (August 1914): 115–22.

———. "The Physique Beautiful—Its Ideal Proportions in Women." *Physical Culture* 34, 3 (September 1915): 6–35.

Hawkes, Kristen. "Showing Off: Tests of Another Hypothesis about Men's Foraging Goals." *Ethology and Sociobiology* 12, 1 (1991): 29–54.

Hess, Eckard H. "Comparative Psychology." In Calvin Perry Stone and Donald Wayne Taylor, eds. *Annual Review of Psychology.* Vol. 4. Stanford, Calif.: Annual Reviews, 1953.

Hirsch, Jerry. "Behavior-Genetics, or 'Experimental' Analysis: The Challenge of Science versus the Lure of Technology." *American Psychologist* 22 (February 1967): 118–30.

Hogben, Lancelot. *Genetic Principles in Medicine and Social Science.* New York: Alfred A. Knopf, 1932.

Holmes, S.J. "The Role of Sex in the Evolution of the Mind." *The Popular Science Monthly* (August 1914): 200–3.

———. *The Trend of the Race: A Study of Present Tendencies in the Biological Development of Civilized Mankind.* New York: Harcourt, Brace and Co., 1921.

———. *Studies in Evolution and Eugenics.* New York: Harcourt, Brace and Co., 1923.

———. "Evolution and the Ethical Ideal." *The University of California Chronicle* (January 1924): 5–26.

———. "Nature versus Nurture in the Development of the Mind." *The Scientific Monthly* 31, 3, (September 1930): 245–52.

———. "The Opposition to Eugenics." Presidential Address before the American Eugenics Society, New York, November 30, 1938. Reprinted in *Science* 89, 2312 (April 21, 1939): 351–7.

———. "Darwinian Ethics and Its Practical Applications." *Science* (New Series) 90, 2328 (August 11, 1939): 117–23.

———. *Life and Morals.* New York: The Macmillan Company, 1948.

Holmes, S.J. and C.E. Hatch. "Personal Appearance as Related to Scholastic Records and Marriage Selection in College Women." *Human Biology* 10, 1 (February 1938): 65–76.

Horgan, J. "Eugenics Revisited: Trends in Behavioral Genetics." *Scientific American* 268 (1993): 122–31.

———. "The New Social Darwinists." *Scientific American* 273 (1995): 174–81.

Howard, G.E. *History of Matrimonial Institutions.* Vol. I. New York: Humanities Press, 1964 [1904].

Hunt, Harrison Randall. "Matrimonial Views of University Students." *Journal of Heredity* (1921): 14–21.

———. *Some Biological Aspects of War.* New York: Galton Publishing Company, 1930.

Huxley, Julian. "Ils n'ont que de l'âme: an essay on bird-mind." In Julian Huxley. *Essays of a Biologist.* London: Chatto & Windus, 1923.

———. "Progress, Biological and Other." In Julian Huxley. *Essays of a Biologist*. London: Chatto & Windus, 1923.

———. "The Uniqueness of Man." *Yale Review* (New Series) 28 (1938): 473–500.

———. *Evolution: The Modern Synthesis*. New York, London: Harper & Bros., 1942.

———. "Introduction." In "A Discussion on Ritualization of Behaviour in Animals and Man." *Philosophical Transaction of the Royal Society of London. Series B. Biological Sciences* 251, 772 (December 29, 1966): 249–71.

———. *The Courtship Habits of the Great Crested Grebe: With an Addition to the Theory of Sexual Selection*. London: Jonathan Cape, 1968 [1914].

James, William. *The Principles of Psychology*. New York: H. Holt and Company, 1890. http://psychclassics.yorku.ca/James/Principles/prin24.htm. Accessed April 4, 2004.

Jennings, Herbert Spencer. *Prometheus; or Biology and the Advancement of Man*. New York: E.P. Dutton & Company, 1925.

Johnson, Roswell H. "The Evolution of Man and Its Control." *The Popular Science Monthly* (January 1910): 49–70.

———. "The Improvement of Sexual Selection." In Emily F. Robbins, ed. *Proceedings of the First National Conference on Race Betterment Battle Creek, MI. (Jan. 8–12, 1914)*. Battle Creek, Mich.: Gage Printing Co., 1914, 515–32.

———. "Mate Selection." *The Eugenics Review* 14, 4 (January 1923): 258–65.

Jones, Doug. "Sexual Selection, Physical Attractiveness, and Facial Neoteny." *Current Anthropology* 36, 5 (December 1995): 723–48.

Judd, C.H. "Evolution and Consciousness." *The Psychological Review* (New Series) 17, 2 (March 1910): 77–97.

Kaempffert, Waldemar. "Year's Review of the Sciences Reveals Much New Knowledge." *New York Times* (January 3, 1932): XX5.

———. "Review of Lancelot Hogben. *Genetic Principles in Medicine and Social Science*." *New York Times* (September 25, 1932): BR17.

———. "New Horizons for Science." *New York Times* (September 2, 1934): BR1.

Kaye, S.A., A.R. Folsom, R.J. Prineas, J.D. Potter, and S.M. Gapstur. "The Association of Body Fat Distribution with Lifestyle and Reproductive Factors in a Population Study of Postmenopausal Women." *International Journal of Obesity* 14 (1990): 583–91.

Köhler, Wolfgang. *The Mentality of Apes*. London: Routledge, 1999 [1925].

Kroeber, Alfred L. "Inheritance by Magic." *American Anthropologist* (New Series) 18 (1916): 19–40.

———. "The Superorganic." *American Anthropologist* 19 (1917): 169.

Kuhn Jr., Ferdinand. "Forum on Behavior Most Misbehaved." *New York Times* (September 12, 1934): 10.

Kuo, Zing Yang. "Giving Up Instincts in Psychology." *Journal of Philosophy* 18, 24 (November 24, 1921): 645–64.

———. "The Net Result of the Anti-Heredity Movement in Psychology." *The Psychological Review* 36, 3 (May 1929): 181–99.

Lang, Andrew and J.J. Atkinson. *Social Origins*. London, New York, and Bombay: Longmans, Green, and Co., 1903.

Langlois, Judith and Lori Roggman. "Attractive Faces Are only Average." *Psychological Sciences* 1, 2 (March 1990): 115–21.

Lashley, K.S. "Experimental Analysis of Instinctive Behavior." *Psychological Review* 45 (1938): 443–71.

Laurence, William L. "Sees a Super-Race Evolved by Science." *New York Times* (August 25, 1932): 40.

Lenz, Fritz. *Menschliche Auslese und Rassenhygiene*. Munich: J.F. Lehmann, 1931.

Lippman, Walter. "The Mental Age of Americans." *New Republic* 32, 412 (October 25, 1922): 213–15; no. 413 (November 1, 1922): 246–8; no. 414 (November 8, 1922): 275–7; no. 415 (November 15, 1922): 297–8; no. 416 (November 22, 1922): 328–30; no. 417 (November 29, 1922): 9–11.

Lord Raglan, "Incest and Exogamy." *The Journal of the Royal Anthropological Institute of Great Britain and Ireland* 61 (January–June 1931): 167–80.

Lorenz, Konrad. *On Aggression*. Trans. Marjorie Kerr Wilson. New York: Harcourt, Brace & World, Inc., 1963.

———. *The Foundations of Ethology*. Trans Konrad Z. Lorenz and Robert Warren Kickert. New York: Springer-Verlag, 1981.

Macfadden, Bernarr. *Manhood and Marriage*. New York: Physical Culture Publishing Co., 1916.

Malcolm, Marion. "What Is the Ideal Female Form?" *Physical Culture* 31, 5 (May 1914): 439–45.

———. "Are Women Growing Taller?" *Physical Culture* 33, 2 (February 1915): 158–62.

Malinowski, Bronislaw. *Sex and Repression in Savage Society*. London: Routledge & Kegan Paul, 1927.

Manning, J.T., D. Scutt, G.H. Whitehouse, S.J. Leinster, and J.M. Walton. "Asymmetry and the Menstrual Cycle in Women." *Ethology and Sociobiology* 17 (1996): 129–30.

McCurdy, Harold G. "William McDougall." In Benjamin B. Wolman, ed. *Historical Roots of Contemporary Psychology*. New York: Harper and Row, 1968.

McDougall, William. "Motives in the Light of Recent Discussion." *Mind* 29 (1920): 277–93.

———. *Is America Safe for Democracy?* New York: Scribner, 1921.

———. "The Use and Abuse of Instinct in Social Psychology." *The Journal of Abnormal and Social Psychology* 16 (1922): 285–333.

———. *Introduction to Social Psychology*. Revised ed. Boston: John W. Luce & Co., 1926.

———. "Objection and Reproof." *The New York Times* (May 27, 1928): 68.

———. *Introduction to Social Psychology*. 14th ed. Kitchener, Ont.: Batoche Books, 2001 [1919].

McDougall, William with J.B. Watson. *The Battle of Behaviorism: An Exposition and an Exposure*. New York: W.W. Norton & Co., 1929.

Miller, Geoffrey. "Evolution of the Human Brain through Runaway Sexual Selection: The Mind as a Protean Courtship Device." Unpublished thesis, 1994. http://www.serpentfd.org/a/miller1994.html. Accessed November 1, 2003.

———. "Some Elements of Sexual Behavior in Primates and Their possible Influence on the Beginnings of Human Social Development." *Journal of Mammalogy* 9, 4 (November 1928): 284–90.

————. "Aesthetic Fitness: How Sexual Selection Shaped Artistic Virtuosity as a Fitness Indicator and Aesthetic Preferences as Mate Choice Criteria." *Bulletin of Psychology and the Arts* 2, 1 (2000): 20–5.

————. "Evolution of Human Music through Sexual Selection." In N.L. Wallin, B. Merker, and S. Brown, eds. *The Origins of Music*. Cambridge: MIT Press, 2000.

Miller, Jr., Gerrit S. "The Primate Basis of Human Sexual Behavior." *The Quarterly Review of Biology* 6, 4 (December 1931): 379–410.

Montagu, Ashley. "The New Litany of 'Innate Depravity,' or Original Sin Revisited." In Ashley Montagu, ed. *Man and Aggression*. New York: Oxford University Press, 1968.

Morgan, Thomas Hunt. *Evolution and Genetics*. Princeton: Princeton University Press, 1925.

Morris, Desmond. "The Rigidification of Behaviour." In "A Discussion on Ritualization of Behaviours in Animals and Man." *Philosophical Transaction of the Royal Society of London. Series B, Biological Sciences* 251, 772, (December 29, 1966).

————. *The Naked Ape: A Zoologist's Study of the Human Animal*. New York: McGraw-Hill, 1967.

————. Introduction to Julian Huxley, *The Courtship Habits of the Great Crested Grebe: With an Addition to the Theory of Sexual Selection*. London: Jonathan Cape, 1968 [1914].

Moss, Fred August and Edward L. Thorndike. *Comparative Psychology*. New York: Prentice-Hall, 1934.

Muller, Hermann J. "The Dominance of Economics over Eugenics." *A Decade of Progress in Eugenics*. Baltimore: The Williams &Wilkins Company, 1934, 138–44.

Murdock, George. "The Science of Culture." *American Anthropologist* 34, 2 (April–June 1932): 200–14.

Murphy, Gardner. *Historical Introduction to Modern Psychology*. New York: Harcourt, Brace, 1949.

Murray, Henry A. Autobiography. In E.G. Boring and G. Lindzey, eds. *A History of Psychology in Autobiography*. Vol. 5. New York: Appleton-Century-Crofts, 1967.

Naly, Josephine Olivia. "Facial Appearance and Other Factors in Mate Selection among College Graduates." M.S. thesis, University of Pittsburgh, 1929.

Nettle, Daniel. "Women's Height, Reproductive Success and the Evolution of Sexual Dimorphism in Modern Humans." *Proceedings of the Royal Society B: Biological Sciences* 269, 1503 (September 22, 2002): 1919–23.

Nietzsche, Friedrich. *Human, all too Human*. In *The Portable Nietzsche*. Ed. Walter Kaufman, trans. R.J. Hollingdale. New York: The Viking Press, 1954.

Nisbet, J.F. *Marriage and Heredity: A View of Psychological Evolution*. 2nd ed. London: Ward and Downey, 1890.

Nissen, H.W. and M.P. Crawford. "A Preliminary Study of Food-Sharing Behavior in Young Chimpanzees." *Journal of Comparative Psychology* 22 (1936): 383–419.

Ogburn, William Fielding. "Trends in Social Science," *Science* (New Series) 79, 2047 (March 23, 1934): 257–62.

Osborn, Frederick. "The American Concept of Eugenics." *Eugenical News* 24, 3 (March 1939): 2.

Pearl, Raymond. "The Biology of Superiority." *American Mercury* 12 (1927): 257–66.

Penton-Voak, I.S., D.I. Perrett, D.L. Castles, T. Kobayashi, D.M. Burt, L.K. Murray, and R. Minamisawa. "Menstrual Cycle Alters Face Preference." *Nature* 399 (June 24, 1999): 741–2.

Percival, C. Gilbert. "The Woman Beautiful." *Health* 58, 9 (September 1908): 468.

Popenoe, Paul. *Modern Marriage: A Handbook.* New York: Macmillan, 1929.

Popenoe, Paul and Roswell Hill Johnson. *Applied Eugenics.* New York: MacMillan Co., 1918.

Rawlings III, Hunter R. "State of the University Address." October 21, 2005. http://www.cornell.edu/president/announcement_2005_1021.cfm. Accessed November 2, 2006.

Reuter, E.B. "Review of *Applied Eugenics*." *The American Journal of Sociology* 40, 4 (January 1935): 549.

"Review of S.J. Holmes, *The Eugenic Predicament*." *Nation* (January 10, 1934): 51.

"Review of *The Eugenic Predicament*." *Quarterly Review of Biology* 9, 2 (June 1934): 225–58.

Ridley, Matt. *The Red Queen: Sex and the Evolution of Human Nature.* New York: Macmillan Publishing Company, 1993.

Romanes, George John. *Mental Evolution in Man: Origin of Human Faculty.* New York: D. Appleton and Co., 1889.

Ross, Edward A. *The Old World in the New: the Significance of Past and Present Immigration to the American People.* New York: Century Co., 1914.

———. *Social Control: A Survey of the Foundations of Order.* New York: Macmillan Co., 1924.

Ruggeri, Giuffrida. "Our Work in Eugenics." *New York Times* (July 26, 1912): 4.

Schoen, Max. "Instinct and Man." *The Scientific Monthly* 28, 6 (June 1929): 531–8.

Singh, Devendra. "Adaptive Significance of Female Physical Attractiveness: Role of Waist-to-Hip Ratio." *Journal of Personality and Social Psychology* 65, 2 (August 1993): 293–307.

Smith, Miriam Law et al. "Facial Appearance Is a Cue to Oestrogen Levels in Women." *Proceedings of the Royal Society B: Biological Sciences* 273, 1583 (January 22, 2006): 135–40.

Smuts, Barbara. "Sexual Competition and Mate Choice." In Barbara Smuts et al., eds. *Primate Societies.* Chicago: University of Chicago Press, 1986.

Spencer, Susan. "The Importance to Males and Females of Physical Attractiveness, Earning Potential, and Expressiveness in Initial Attraction." *Sex Roles* 21, 9/10 (1989): 591–607.

Stanford, Craig. "The Cultured Ape?" *The Sciences* 40 (May/June 2000): 39.

Symons, Donald. *The Evolution of Human Sexuality.* New York: Oxford University Press, 1979.

Terman, Lewis M. Autobiography. In Carl Murchison, ed. *History of Psychology in Autobiography.* Vol. 2 Worcester, Mass.: Clark University Press, 1930.

Thompson, J. Arthur. "On Sexual Selection." *Scientia* 24, 12 (1918): 31.

Thorndike, Edward L. *Measurement of Twins.* New York: The Science Press, 1905.

———. *Elements of Psychology.* 2nd ed. New York: G. Seiler, 1916.

Thornhill, Randy and Steve Gangestad. "Human Facial Beauty: Averageness, Symmetry, and Parasite Resistance." *Human Nature* 4, 3 (1993): 237–69.

Thorpe. W.H. *The Origins and Rise of Ethology: The Science of the Natural Behaviour of Animals.* New York: Praeger, 1979.

Tilney, Frederick and Henry Alsop Riley. *The Brain from Ape to Man: A Contribution to the Study of the Evolution and Development of the Human Brain.* New York: B.P. Hoeber, 1928.

Tinbergen, Niko. *The Study of Instinct.* New York: Oxford University Press, 1951.

———. "Interview with Niko Tinbergen." In D. Cohen, ed. *Psychologists on Psychology.* New York: Taplinger, 1977.

Tisdale, Hope. "Biology in Sociology." *Social Forces* 18, 1 (October 1939): 29–40.

Trivers, Robert L. "The Evolution of Reciprocal Altruism." *The Quarterly Review of Biology* 46, 1 (March 1971): 35–57.

———. "Parental Investment and Sexual Selection." In B. Campbell, ed. *Sexual Selection and the Descent of Man 1871–1971.* Chicago: Aldine, 1972.

Trotter, William. *Instincts of the Herd in Peace and War.* London: T. Fisher Uwin Ltd., 1919 [1915].

Udry, J.R. and Bruce K. Eckland. "Benefits of Being Attractive: Differential Payoffs for Men and Women." *Psychological Reports* 54 (1984): 47–56.

Walters, Sally and Charles B. Crawford. "The Importance of Mate Attraction for Intrasexual Competition in Men and Women." *Ethology & Sociobiology* 15, 1 (January 1994): 5–30.

Ward, Lester. *Outlines of Sociology.* New York: MacMillan, 1899.

Watson, John B. *Behaviorism.* Chicago: University of Chicago Press, 1958 [1924].

———. *Behaviorism.* Revised ed. Chicago: University of Chicago Press, 1930.

Westermarck, Edward. *The History of Human Marriage.* Vol. I, 5th ed. New York: Allerton Book Co., 1922.

———. "Methods in Social Anthropology." *The Journal of the Royal Anthropological Institute of Great Britain and Ireland* 66 (July–December 1936): 223–38.

Williams, Carl Easton. "Womanly Beauty among Various Races." *Physical Culture* 29, 6 (June 1913): 533–40.

———. "Do Men Choose Wives for Beauty?" *Physical Culture* 33, 1 (January 1915): 13–20.

Wilson, Edmund O. *Sociobiology: The New Synthesis.* Cambridge, Mass.: Belknap Press of Harvard University Press, 1975.

———. "Academic Vigilantism and the Political Significance of Sociobiology." *BioScience* 26 (1976): 187–90.

———. *On Human Nature.* Cambridge, Mass.: Harvard University Press, 1978.

———. *Naturalist.* Washington, D.C.: Island Press, 1994.

Witty, Paul A. and Harvey C. Lehman. "The Dogma and Biology of Human Inheritance." *The American Journal of Sociology* 35, 4 (January 1930): 548–63.

Yerkes, Robert M. "The Study of Human Behavior." *Science* (New Series) 39, 1009 (May 1, 1914): 625–33.

———. "Recent Psycho-Biological Studies of the Chimpanzee." *Eugenic News* 10, 4 (April 1925): 65–6.

Yerkes, Robert M. *Almost Human*. New York and London: The Century Co., 1925.

———. "Mental Evolution in the Primates." In E.V. Cowdry, ed. *Human Biology and Racial Welfare*. New York: Paul B. Hoeber, Inc., 1930.

———. "Genetic Aspects of Grooming, a Socially Important Primate Behavior Pattern." *The Journal of Social Psychology* 4, 1 (February 1933): 3–26.

———. "The Significance of Chimpanzee-Culture for Biological Research." *Harvey Lecture Series* (1935): 57–73.

———. "Primate Cooperation and Intelligence." In Karl Dallenbach, ed. *The American Journal of Psychology, Golden Jubilee Volume L*. Ithaca, N.Y.: Cornell University Press, 1937.

———. "The Method and Scope of Science." *Endeavor* 5 (1946): 126.

———. "The Scope of Science." *Science* (New Series) 105, 2731 (May 2, 1947): 461–3.

Zaadstra, B. et al. "Fat and Female Fecundity: Prospective Study of Effect of Body Fat Distribution on Conception Rates." *British Medical Journal* 306 (1993): 484–7.

Zahavi, Amotz. "Mate Selection—A Selection of Handicap." *Journal of Theoretical Biology* 53 (1975): 207–14.

Secondary Sources

"Eroticism and Gender." *U.S. News and World Report* (July 19, 1993): 62.

"Recent Psycho-Biological Studies of the Chimpanzee." *Eugenic News* 10, 4 (April 1925): 65–6.

"Relationships." *Psychology Today* (May/June 2000): 16.

"Samuel Holmes, Zoologist, Dead." *New York Times* (March 8, 1964): 87.

Alcock, John. *The Triumph of Sociobiology*. Oxford: Oxford University Press, 2001.

Allman, William F. "The Mating Game." *U.S. News and World Report* (July 19, 1993): 56–63.

Alroy, John. "Lefalophodon: An Informal History of Evolutionary Biology Web Site." http://www.nceas.ucsb.edu/people/alroy/public_html/lefa/Conklin.html. Assessed November 1, 2006.

Bachman, J. "Theodore Lothrop Stoddard: The Bio-Sociological Battle for Civilization." Ph.D. thesis, University of Rochester, 1966.

Badcock, Christopher. *Evolutionary Psychology: A Critical Introduction*. Cambridge, Eng.: Polity, 2000.

Baker, R. Robin and Mark A. Bellis. *Human Sperm Competition: Copulation, Masturbation, and Infidelity*. London: Chapman and Hall, 1995.

Barber, N. "The Evolutionary Psychology of Physical Attractiveness: Sexual Selection and Human Morphology." *Ethology and Sociobiology* 16 (1995): 395–424.

Barkan, E. *The Retreat of Scientific Racism: Changing Concepts of Race in Britain and the United States between the World Wars*. Cambridge: Cambridge University Press, 1992.

Barkow, Jerome, Leda Cosmides, and John Tooby, eds. *The Adapted Mind: Evolutionary Psychology and the Generation of Culture*. New York: Oxford University Press, 1992.

Begley, Sharon. "You Must Remember This, a Kiss Is but a Kiss: Infidelity and the Science of Cheating." *Newsweek* (December 30, 1996): 59–63.

———. "Boys Need Not Be Boys." *Newsweek* (April 12, 1999): 69.

Bergman, Jerry. "A Brief History of the Eugenics Movement." *Investigator* 72 (May 2000). http://www.adam.com.au/bstett/BEugenics72Bergman73Potter77.htm. Accessed June 13, 2004.

Blustain, Sarah. "The New Gender Wars." *Psychology Today* (November/December 2000): 42–9.

Boakes, Robert. *From Darwin to Behaviorism: Psychology and the Minds of Animals*. New York, London: Cambridge University Press, 1984.

Bowler, Peter J. *Evolution: The History of an Idea*. 3rd ed. Berkeley and Los Angeles: University of California Press, 2003.

Buss, Allan R., ed. *Psychology in Social Context*. New York: Irvington Publishers, 1979.

Buss, David M. *The Evolution of Desire: Strategies of Human Mating*. New York: Basic Books, 1994.

———. *Evolutionary Psychology: The New Science of the Mind*. Needham Heights, Mass.: Allyn and Bacon, 1999.

Campbell, Anne. *A Mind of Her Own: The Evolutionary Psychology of Women*. Oxford: Oxford University Press, 2002.

Carpenter, Betsy. "Nature's Son." *U.S. News and World Report* (October 21, 2002): 70–1.

Cartwright, John. *Evolution and Human Behavior: Darwinian Perspectives on Human Nature*. Cambridge: MIT press, 2000.

Cowley, Geoffrey. "The Biology of Beauty." *Newsweek* 127, 23 (June 3, 1996): 60–7.

Cravens, Hamilton. *The Triumph of Evolution: The Heredity–Environment Controversy, 1900–1941*. Baltimore: Johns Hopkins, 1988 [1978].

Crawford, Charles, Martin Smith, and Dennis Krebs, eds. *Sociobiology and Psychology: Ideas, Issues and Applications*. Hillsdale, N.J.: Lawrence Erlbaum Associates, 1987.

De Grazia, Victoria. *How Fascism Ruled Women: Italy, 1922–1945*. Berkeley: University of California Press, 1993.

de Waal, Frans. *Our Inner Ape: A Leading Primatologist Explains Why We Are Who We Are*. New York: Riverhead Books, 2005.

Degler, Carl N. *In Search of Human Nature: The Decline and Revival of Darwinism in American Social Thought*. New York: Oxford University Press, 1991.

DeVore, Irven. *Primate Behavior: Field Studies of Monkeys and Apes*. New York: Holt, Rinehart and Winston, 1965.

Dokötter, Frank. "Race Culture: Recent Perspectives on the History of Eugenics." *American Historical Review* 103 (1998): 467–78.

Dunbar, Robert, Louise Barrett, and John Lycett. *Evolutionary Psychology: A Beginner's Guide*. Oxford: Oneworld Publications, 2005.

Eakin, Richard M., Charles L. Camp, and Kenneth DeOme. "1965, University of California: In Memoriam. Samuel Jackson Holmes, Zoology: Berkeley."

http://dynaweb.oac.cdlib.org:8088/dynaweb/uchist/public/inmemoriam/inmemoriam1965/@Generic__BookTextView/502. Accessed June 21, 2004.

Elliot, Richard. "Robert Mearns Yerkes: 1876–1956." *American Journal of Psychology* 69, 3 (September 1956): 487–9.

Fischman, Josh. "Why We Fall in Love." *U.S. News and World Report* (February 7, 2000): 42–9.

Flynn, Emily. "Beauty: Babes Spot Babes." *Newsweek* (September 20, 2004): 10.

Garcia, Agnaldo. "The Psychological Literature in Konrad Lorenz's Work: A Contribution to the History of Ethology and Psychology." *Memorandum* 5 (2003): 105–33. http://www.fafich.ufmg.br/~memorandum/artigos05/garcia01.htm.

Gillette, Aaron. *The History of Latin Eugenics* (manuscript).

Gorman, Christine. "Sizing Up the Sexes: Scientists Are Discovering that Gender Differences Have as much to Do with the Biology of the Brain as the Way We Are Raised." *Time Magazine* (January 20, 1992): 38–45.

Graham, Richard. *The Idea of Race in Latin America, 1870–1940.* Texas: University of Texas Press, 1990.

Haller, Mark. *Eugenics: Hereditarian Attitudes in American Thought.* New Brunswick, N.J.: Rutgers University Press, 1984.

Haslerud, George M. Preface to Robert Yerkes. *The Mental Life of Monkeys and Apes.* Demar, N.Y.: Scholars; Facsimiles and Reprints, 1979 [1916].

Herrnstein, R.J. "Nature as Nurture: Behaviorism and the Instinct Doctrine." *Behaviorism* 1 (1972): 23–52.

Holowchak, Mark Andrew. "Critical Reasoning & Science: Thinking about Science with a Critical Eye" (unpublished manuscript).

Holstein, Jean and William L. Dupuy. *The First Fifty Years at the Jackson Laboratory.* Bar Harbor, Maine: Jackson Laboratory, 1979.

Horn, David G. "Constructing the Sterile City: Pronatalism and Social Sciences in Interwar Italy." *American Ethnologist* 18, 3 (August 1991): 581–601.

Israel, Giorgio and Pietro Nastasi. *Scienza e razza nell'Italia fascista.* Bologna: Il Mulino, 1998.

Jackson, L.A. *Physical Appearance and Gender: Sociobiological and Sociocultural Perspectives.* Albany: SUNY Press, 1992.

Jones, Russell. "Psychology, History, and the Press: The Case of William McDougall and *The New York Times.*" *American Psychologist* 42, 10 (October 1987): 931–40.

Kandel, Eric R. "A New Intellectual Framework for Psychiatry." *American Journal of Psychiatry* 15, 5 (April 1998): 457–69.

Katcher, Brian S. "The Post-Repeal Eclipse in Knowledge about the Harmful Effects of Alcohol." *Addiction* 88 (1993): 729–44.

Kevles, Daniel J. and Leroy Hood, eds. *The Code of Codes: Scientific and Social Issues in the Human Genome Project.* Cambridge: Harvard University Press, 1992.

Kluger, Jeffrey, Jeff Chu, Broward Liston, Maggie Sieger, and Daniel Williams. "Is God in Our Genes?," *Time* (October 25, 2004): 62.

———. "Ambition: Why Some People Are Most Likely to Succeed." *Time* (November 14, 2005): 48.

Krantz, David L. and David Allen. "The Rise and Fall of McDougall's Instinct Doctrine." *Journal of the History of the Behavioral Sciences* 3 (1967): 326–38.

Kühl, Stefan. *The Nazi Connection: Eugenics, American Racism, and German National Socialism.* New York: Oxford University Press 1994.

———. *Die Internationale der Rassisten: Aufstieg und Niedergang der internationalen Bewegung für Eugenik und Rassenhygiene im 20. Jahrhundert.* Frankfurt am Main: Campus, 1997.

Ladd-Taylor, Molly. "Eugenics, Sterilisation and the Modern Marriage in the USA: The Strange Career of Paul Popenoe." *Gender & History* 13, 2 (August 2001): 298–327.

Lemonick, Michael D., Dan Cray, and Wendy Grossman. "Honor among Beasts." *Time* (July 11, 2005): 54.

Ludmerer, Kenneth. *Genetics and American Society: A Historical Appraisal.* Baltimore: The Johns Hopkins University Press, 1972.

Ludovici, Anthony M. *The Choice of a Mate.* London: John Lane The Bodley Head, 1935.

Marks, Jonathan. "Introduction: Non-, Para-, and Anti-Boasians." http://www.aaanet.org/gad/history/Intro09.pdf. Accessed November 1, 2006.

Marquis, Albert Nelson, ed. *Who's Who in America 1950–1951.* Vol. 26. Chicago: The A.N. Marquis Co., 1950–1951.

McDougall, William. *Group Mind.* New York, London: G.P. Putnam's Sons, 1920.

Mehler, Barry Alan. "A History of the American Eugenics Society, 1921–1940." Ann Arbor, MI.: UMI, 1988.

Miller, Geoffrey F. "How Mate Choice Shaped Human Nature: A Review of Sexual Selection and Human Evolution." In C. Crawford and D. Krebs, eds. *Handbook of Evolutionary Psychology: Ideas, Issues, and Applications.* Mahwah, N.J.: Lawrence Erlbaum, 1998.

———. *The Mating Mind: How Sexual Choice Shaped Human Nature.* New York: Doubleday, 2000.

———. "A Review of Sexual Selection and Human Evolution: How Mate Choice Shaped Human Nature." http://www.unm.edu/~psych/faculty/mate_choice.htm. Accessed December 6, 2003.

Miller, Marvin. *Terminating the "Socially Inadequate": The American Eugenicists and the German Race Hygienists, California to Cold Spring Harbor, Long Island to Germany.* New York: Malamud-Rose Commack, 1996.

Murray, H.A. Autobiography. In E.G. Boring and G. Lindzey, eds. *A History of Psychology in Autobiography.* Vol. 5 New York: Appleton-Century-Crofts, 1967.

National Academy of Sciences. "Clarence C. Little." *Biographical Memoirs* 46 (1975): 248.

Newbold, Sally Jean. "William McDougall, MD, FRS, 'A Psychologist in Changing Times.'" Honors Essay, Department of History, University of North Carolina, 1979.

Obituary: "Milo Hastings, 72, an Ex-Food Editor." *New York Times* (February 26, 1957): 29.

Ostrow, Laysha. "Do We Overestimate Our Own Desirability?" *Psychology Today* (September/October 2002): 24.

Paul, Annie Murphy. "Blinded by Beauty." *Psychology Today* (May/June 1998): 17.

Paul, Diane. "Eugenics and the Left." *Journal of the History of Ideas* 45, 4 (October–December 1984): 567–90.

Pauly, Philip J. "How Did the Effects of Alcohol on Reproduction Become Scientifically Uninteresting?" *Journal of the History of Biology* 29 (1996): 1–28.

———. *Biologists and the Promise of American Life: From Meriwether Lewis to Alfred Kinsey.* Princeton: Princeton University Press, 2000.

Pearson, Karl. *The Life, Letters and Labours of Francis Galton.* Vol. IIIa: *Correlation, Personal Identification and Eugenics.* Cambridge: Cambridge University Press, 1930.

Pinker, Steven. *The Blank Slate: The Modern Denial of Human Nature.* New York: Viking, 2002.

Plotkin, Henry. *Evolutionary Thought in Psychology: A Brief History.* Malden, Mass.: Blackwell Publishing, 2004.

Proctor, Robert. *The Nazi War on Cancer.* Princeton, N.J.: Princeton University Press, 1999.

———. "The Nazi Campaign against Tobacco: Science in a Totalitarian State." In Francis R. Nicosia and Jonathan Huener, eds. *Medicine and Medical Ethics in Nazi Germany: Origins, Practices, Legacies.* New York; Oxford: Berghahn Books, 2002.

Rader, Karen. "C.C. Little and the Jackson Laboratory Archives: Some Notes on the Intersecting Histories of Eugenics, Mammalian Genetics, and Cancer Research." *Mendel Newsletter* (New Series) 5 (February 1996). http://www.amphilsoc.org/library/mendel/1996.htm. Accessed June 12, 2004.

"Review of *Eugenical News* Robert M. Yerkes and Blanche W. Learned, *Chimpanzee Intelligence and Its Vocal Expression*." *Eugenical News* 10, 9 (September 1925): 124–5.

Roback, A.A. *A History of American Psychology.* Rev. ed.. New York: Collier, 1964.

Roll-Hansen, Nils. "Eugenics before World War II: The Case of Norway." *History and Philosophy of the Life Sciences* 2, 2 (1980): 267–98.

Sachs, Andrea. "Dangerous Steps." *Time Select* (November 15, 1999).

Samelson, Franz. "Putting Psychology on the Map: Ideology and Intelligence Testing." In Allan R. Buss, ed. *Psychology in Social Context.* New York: Irvington Publishers, 1979.

Schmitt, David P. "Fundamentals of Human Mating Strategies." In David M. Buss, ed. *The Handbook of Evolutionary Psychology.* Hoboken, N.J.: John Wiley & Sons, 2005.

Slater, Lauren. "Why We Have Children." *Parenting* 19, 11 (December 2005–January 2006): 192.

Spiro, Jonathan P. "Nordic vs. Anti-Nordic: The Galton Society and the American Anthropological Association." *Patterns of Prejudice* 36, 1 (2002): 35–48.

Stein, Rob. "Common Collie or Uberpooch?," *The Washington Post* (June 11, 2004): A01.

Stocking, Jr., George W. *Race, Culture, and Evolution: Essays in the History of Anthropology.* Chicago: University of Chicago Press, 1982 [1968].

———, ed. *Bones, Bodies, Behavior: Essays on Biological Anthropology.* Madison: University of Wisconsin Press, 1988.

Thornhill, Randy and Palmer, Craig T. *A Natural History of Rape: Biological Basis of Sexual Coercion.* Cambridge, Mass.: MIT Press, 2000.

Todd, Jan. "Bernarr Macfadden: Reformer of Feminine Form." *Journal of Sport History* 14, 1 (Spring 1987): 61–74. http://www.aafla.org/SportsLibrary/JSH/JSH1987/JSH1401/jsh1401e.pdf. Accessed July 20, 2004.

Tooby, John. "Jungle Fever: Did Two U.S. Scientists Start a Genocidal Epidemic in the Amazon, or was the *New Yorker* Duped?" *Slate* (October 25, 2000). http://www.slate.com/id/91946/. Accessed October 30, 2006.

Tucker, William H. *The Science and Politics of Racial Research.* Bloomington, IL: University of Illinois Press, 1994.

———. *The Funding of Scientific Racism: Wickliffe Draper and the Pioneer Fund.* Chicago: University of Illinois Press, 2002.

Turner, Oliver. "Our Cheating Hearts." *Psychology Today* (November/December, 2000): 17.

Wilkinson, Joseph F. "Look at Me." *Smithsonian* 28, 9 (December 1997): 136–47.

Wispé, Lauren G. and James. N. Thompson, Jr. "The War between the Words: Biological versus Social Evolution and Some Related Issues." *American Psychologist* 31, 5 (May 1976): 341–8.

Wolman, Benjamin B., ed. *Historical Roots of Contemporary Psychology.* New York: Harper and Row, 1968.

Wright, Robert. *The Moral Animal: The New Science of Evolutionary Psychology.* New York: Pantheon Books, 1994.

———. "Dancing to Evolution's Tune." *Time* (January 17, 2005): A11.

Index

1920s 113, 114, 121
Advisory Committee on Research on
    Human Behavior, Carnegie
    Institute of Washington 54–5,
    120
Aggression, instinctual 21, 35–6,
    43–5, 161
Agnatology 16
Alcock, John 36
Alcohol, health hazards, research
    on 17
Allee, Warder Clyde 97, 159
Allen, David 152
Allen, William 154
Altruism
    definition of 173 n. 6
    group 96, 99
    instinct and 22, 36–7, 95–100,
        177 n. 26
Alverdes, Friedrich 41, 46, 49, 95, 97,
    107, 177 n. 26–7, 188–9 n. 1
American Anthropological Association
    110, 112, 119
American Association for the
    Advancement of Science 1, 60,
    180 n. 81, 181 n. 22
American Association of Physical
    Anthropologists 162
American Cancer Society 158
American Eugenics Society 12, 53,
    56, 60, 68, 78, 119, 126, 138–9,
    142, 143, 136–7, 146, 149, 151,
    158, 177 n. 12, 178 n. 55, 196 n. 47
American Psychological Association
    22, 51, 116, 118, 127, 129, 134, 166

American Society for Human Genetics
    154
American Sociological Review 166
American Sociological Society 151
Annual Review of Psychology 165–6
Anthropology and Prehistoric
    Archeology, Seventh International
    Congress of 132
Anti-Semitism 140, 142, 144
Art 101, 160
Atkinson, J.J. 93

Baker, Robin 94
Barnes, Michael 34, 87
Beach, Frank 165
Beauty, physical 23, 30–3, 61–94,
    182 n. 53, 184 n. 82, 86, 185 n. 114,
    186 n. 139
Bees 22–3
Behaviorism, definition of 12
Bellis, Mark 94
Benedict, Ruth 110–1
Bigelow, Maurice 196 n. 47
Binder, Rudolph M. 85, 88
Bingham, Harold 54, 57, 101,
    180 n. 76
Birth Control 11, 71, 79, 137, 138,
    194 n. 14
Birth Control Conference, First
    American 137–8
Blough, Donald S. 165–6
Boakes, Robert 111
Boas, Franz 59, 109–10, 112, 114,
    120, 126, 133, 135, 142, 145,
    189 n. 12

Briffault, Robert   131
Brigham, Carl   51–2, 137
British Association for the
    Advancement of Science   129,
    150
Bush, Vannevar   143
Buss, David   14, 28, 34, 87
Butler, Nicholas   112

California Commonwealth Club   151
California, University of   12, 22, 60,
    76, 82, 83
Calverton, V.F.   131
Campbell, Anne   35, 91
Campbell, Clarence G.   108, 142,
    192 n. 17
Carmichael, Leonard   166, 192 n. 22
Carnegie Institute of Washington   11,
    53–5, 76, 120, 127, 136–7, 143
Castle, William   123, 124, 157–8
Cattell, James   127
Chagnon, Napoleon   14
Chicago, University of   76, 82
Chimpanzee and human behavior,
    comparison of   25, 31, **49–60**,
    89–90, 97, 101, 126, 129–30, 148,
    180 n. 74, 76
Chimpanzees, culture and   25, 27
Clark University   42, 75, 109
Clark, Edwin L.   85, 88
Cold War   157
Columbia University   11, 12, 76, 110,
    112, 153
Commission for the Study of
    Population, Italian   79, 82
Conklin, Edwin G.   11, 53, 123, 137,
    138
Coon, Carleton   147, 162
Craig, Wallace   45
Crawford, Charles Bates   35
Crawford, M.P.   98–9
Crowther, J.G.   133
Culture, universal traits of human
    14–15, 67, 111

Darwin, Charles   42, 47, 61, 67, 89,
    90, 91

Darwin, Leonard   64, 84, 150–1
Darwinian evolutionary theory
    8–10, 41–2, 62, 71, 166,
    171–2 n. 5
    and natural selection   8–10, 24,
    74, 96
    and sexual selection   8–10, 16,
    23–4, 28–34, **61–94**, 103
    and sexual selection, criticism of,
    63, 85
Davenport, Charles   11, 49, 52, 55, 65,
    75–6, 77, 108, 112, 123, 126, 133,
    135, 137, 141, 142, 143, 154, 158,
    178 n. 55
Davey, John   61
Dearborn, Walter   192 n. 22
Democrats, New Deal   12, 125, 140,
    143
Depression, Great   108, 121, 125, 126,
    135, 137
DeVore, Irven   37
Drew, Frank   75
Duffus, R.L.   129
Dunlap, Knight   44, 66, 68–9, 87,
    122–3, 147, 177 n. 12,
    181 n. 22, 29
Dunn, Leslie   137, 154, 197 n. 83

East, Edward M.   1, 123, 138
Eaton, R.L.   167
Eckland, Bruce K.   33, 80, 166
Ectoff, Nancy   31
Eibl-Eibesfeldt, Irenäus   72
Ellis, Havelock   69, 70, 71, 73, 92,
    181 n. 111
Ellwood, Charles A.   41, 44, 99, 101
Emlen, Stephen T.   36
Endeavor   160
Environmental behaviorism,
    definition of   171 n. 4
Ethology   44, 45, 161, 164, 165, 167
Eugenical News   56, 142, 145,
    192 n. 17
Eugenics Committee of the United
    States   138
Eugenics Record Office   11, 50, 123,
    136–7, 143, 145, 153

Eugenics Research Association 85, 125, 141, 144–5, 192 n. 17
Eugenics 136
Eugenics
 definition of 2, 10, 63
 democracy and 85, 116, **143–6**
 Latin 79
 Nazi Germany and 1, 13, 16, 108, 135, 139, 141, 142, 144–5, 146, 147, 155
 sterilization and 1, 64, 68, 77, 79, 133, 140, 141, 144, 147, 151, 158
Eugenics, Third International Congress of 137
Evolutionary psychology, definition of 171–2 n. 5

Fairchild, Henry Pratt 11, 56, 118, 119, 138
Feminists 18
Féré, Charles 74
Ferrero, Guigliemo 79, 86, 91
Fisher, Ronald Aylmer 24, 67
Fol, Hermann 70, 182 n. 38
Fort, William, Jr. 134
Freeman, Derek 164

Gallichan, Walter 74, 87
Galton Society 11, 52, 53, 56, 57, 58, 85, 112, 135–6, 139, 140, 142–3, 158
Galton, Francis 10, 63, 67, 99, 117, 141, 176 n. 6
Gangestad, Steve 32
Garcia, Agnaldo 164
Garcia, John 22
Genetics and Social Behavior Symposium 159
Genetics Manifesto 146
Genetics 65, 83, 122, 123, 130, 133, 137, 141, 148, 153, 154, 159, 166, 172 n. 5
Genetics, Sixth International Congress of 158
Genus 161
Gesell, Arnold 92

Gifted and Talented Programs 151–2
Gillett Bill 138
Gilmore, Carrie F. 77–9, 82, 183 n. 78–9
Gini, Corrado 79, 82, 137, 159, 161, 189 n. 12
Goddard, Henry H. 50, 51
Goethe, Charles Matthias 85, 151, 152
Goodall, Jane 25
Gould, Stephen Jay 1
Grammer, Karl 32
Grant, Madison 11, 52, 64, 112, 125, 126, 136, 143, 145
Graubard, Mark 153
Gregory, William K. 142, 143
Grillenzoni, Carlalberto 70, **80–2**
Groos, Karl 90, 177 n. 27

Haldane, J.B.S. 41, 100, 163, 167
Hall, Granville Stanley 42, 61, 69, 75, 88, 91, 109
Hall, Kenneth Ronald Lambert 37
Hamilton, William 22, 96, 99, 100
Hance, Robert T. 88
Handicap Principle 24
Harlow, Harry 22
Harvard University 1, 12, 44, 49, 50, 115, 116, 123, 127, 149–50, 164, 192 n. 22
Hastings, Milo 71, 73, 74
Hatch, C.E. 83
Hawaii, University of 139
Hawaiian Social Hygiene Society 139
Hawkes, Kristen 28
Hess, Eckhard 165
Hirsch, Jerry 166
Hitler, Adolf 112, 140, 141, 142
Hogben, Lancelot 132, 133
Holmes, Samuel Jackson 1, 46, 60, 82–4, **86–7**, 95, 96, 99, 100, 101, 102, 115–18, 135, 136, 147, 150, 151, 152, 188 n. 22
Hooten, Earnest 142, 143
Hrdlicka, Ales 112, 126, 146, 147
Human Biology 70, 82

Human Genetics, Office of 137
Hunt, Harrison R. 84, 87
Huxley, Julian 44, 67, 115, 131–2, 181 n. 15
Hymenoptera 86, 178 n. 35
Hypatia 6

Immigration restriction 52, 68, 77, 112, 119, 187–8 n. 16
Incest taboo 43, 129–30
Inclusive fitness theory *see also* kinship selection 22–3
Instincts, human 22, 42–4, 46–7, 68, 92, 95, 96, 101, 111, 113, 114, 120, 122, 127, 128, 129, 130, 131, 133, 149, 152, 159, 161, 164, 165, 167, 181 n. 22, 200 n. 8
Institute of Family Relations 139
Intelligence tests 12, 132, 134, 141, 193 n. 42, 198 n. 86
International Federation of Eugenics Organizations 85, 144
Iowa, University of 158
Italy, social sciences and 79

James, William 5, 42, 110, 165, 176 n. 2, 200 n. 8
Jennings, Herbert Spencer 123, 136
Johns Hopkins University 68, 113
Johnson, Roswell Hill 69, 70, 75–9, 84, 87, 88, 91, 100, 101, 137–40, 142
Johnson, Victor 72
Jones, Doug 32
Journal of Animal Behavior 161
Journal of Applied Psychology 152
Journal of Comparative Psychology 161

Kaempffert, Waldemar 132
Kallman, Franz 144–5
Kevles, Daniel 154
Kidder, Alfred Vincent 137, 143
Kinship selection 96, 99, 100, 102
Köhler, Wolfgang 49, 179 n. 61
Krantz, David L. 152

Kroeber, Alfred 110, 111, 164
Kühl, Stefan 11
Kuo, Zing Yang 107, 114

La Cerra, Peggy 33–4
Lang, Andrew 93
Langlois, Judith 67
Lashley, Karl 149–50, 161, 164, 165
Laughlin, Harry 125, 137, 141, 142, 144, 145, 154
Lehman, Harvey C. 132
Lewontin, Richard 1
Lippman, Walter 118
Little, Clarence C. 137, 138, 142, 157–9, 199 n. 6
Lombroso, Cesare 79, 91
London, University of 43, 179 n. 72
Lorenz, Konrad 21, 43, 44, 149, 160, 161, 162, 164, 165, 177–8 n. 34
Lysenko, Trofim 6

MacDowell, E. Carleton 158
Macfadden, Bernarr 71, 74, 137
Maine, University of 158
Malcolm, Marion 73, 182 n. 53
Malinowski, Bronislaw 59, 120, 179 n. 72
Marriage and monogamy 30, 43, 49, 59, 63, 64, 70, 77, 78, 80–9, 129–31
McCurdy, Harold 152, 165, 167
McDougall, William 43–4, 70, 71, 72, 90, 92–3, 113, 114, 115–16, 126–8, 134, 147, 149, 164, 165, 167, 190–1 n. 52, 192 n. 22
McGrew, William 25
Mead, Margaret 110
Merriam, John C. 53–4, 119, 120, 136, 142, 143
Michigan, University of 82, 85, 158, 199 n. 6
Miller, Geoffrey 23–4, 29, 102
Miller, Gerrit, Jr. 58–60, 86, 89, 100–1, 120, 180 n. 74
Millward, Richard B. 165–6
Montagu, Ashley 110, 121, 157, 161

Morgan, Thomas Hunt   123, 136
Morris, Desmond   28, 166, 174 n. 3
Muller, Hermann J.   137, 153
Murdock, George   111, 130
Mussolini, Benito   79, 82

Naly, Josephine Olivia   78–9, 82, 184 n. 86
National Research Council   52, 108, 112, 124–5, 126, 143, 153, 154
Neel, James   14
Neotany, facial   32, 72
New Republic   118
New York Times   83, 129, 132–3, 143
New York University   85
Nisbet, John Ferguson   66, 68, 72, 86, 87
Nissen, Henry W.   98–9, 161–2

Ogburn, William Fielding   151
Ohio State University   85, 88
Osborn, Frederick   64, 124, **140–7**, 154, 155, 196 n. 47
Osborn, Henry Fairfield   124, 140, 142
Oxford University   21, 44, 128

Palama Settlement   139
Palmer, Craig   36, 172 n. 29
Parental Investment theory   23
Patriarchical societies   86, 87–8, 93, 111
Pearl, Raymond   123, 136, 138
Pearson, Karl   96, 176 n. 6, 189 n. 12
Percival, Gilbert C.   70
Perret, David   186 n. 129
Physical Culture   71–4, 84, 87, 88, 185 n. 114
Pinker, Steven   14, 157
Pittsburgh, University of   77, 138–9, 184 n. 82
Platt, Philip S.   139
Polygamy   34, 90, 92, 93, 102
Popenoe, Paul   69, 84, 88, 137–40, 142, 198 n. 1
Popular Science   77, 133

Preferential mating   66–7, 74–5, 83
Primates, non-human   22, 46, 48–60, 83
Proctor, Robert   16
Prostitution   90, 98
Psychoanalysis   154
Psychological Abstracts   152

Racial science   8, 11, 12, 13, 15, 51–2, 109, 116, 122–4
Raglan, Lord   129, 131
Ramos, Domingo   141
Rape   36, 85–6, 96, 100–1, 172 n. 29
Reuter, E.B.   139
Riley, Henry Alsop   57
Rockefeller Foundation   158–9
Roggman, Lori   67
Romanes, Geroge John   147
Roscoe B. Jackson Memorial Laboratory   157–9
Ross, Edward A.   44, 99, 131, 177 n. 12, 187–8 n. 16
Rüdin, Ernst   144
Runaway selection   24, 67

Sanger, Margaret   137, 138
Schizophrenia   144
Schoen, Max   127
Schoolland, John Bernard   152
Science   160
Science
    ideology and   2–7, 9, 13, 17, 58, 60, 65, 103, 107–8, 120, 142, 157, 163–4, 167
    philosophy of   4, 5, 18, 67
Scott, John Paul   159, 161
Seashore, Carl Emil   53, 54
Seville Statement on Violence   21
Sexiness   23, 24, 30–2, **61–94**
Sexual jealousy   35, **91–3**, 111
Showoff Hypothesis   28, 91–2
Singh, Devendra   32, 72, 81
Sinnott, Edmund W.   153
Skinner, B.F.   113, 114, 149, 164
Smithsonian Institution   58, 126

Smoking, health hazards, research on
　　16–17
Smuts, Barbara　31
Snyder, Laurence　153
Social Hygiene Association, American
　　138
Sociobiology, definition of　171 n. 5
Sponsel, Leslie　14
Stanford, Craig　27
Steggerda, Morris　123–4
Sterilization *see*　Eugenics,
　　sterilization
Symons, Donald　23

Terman, Lewis　51, 120, 142
Thompson, J. Arthur　133, 193 n. 48
Thorndike, Edward　44, 53, 117,
　　125, 132, 142, 143, 164, 177 n. 12,
　　193 n. 42
Thornhill, Randy　32, 36, 96,
　　172 n. 29
Thorpe, W.H.　149, 163, 164, 165
Tilney, Frederick　57
Tinbergen, Niko　21, 135, 160, 161,
　　162, 164, 165, 174 n. 3, 177–8 n. 34,
　　199 n. 5
Tisdale, Hope　131, 193 n. 38
Trivers, Robert　23, 36, 89, 97, 173 n. 6
Trotter, William　45, 48, 96, 97
Turner, Terrence　14

Udry, J. Richard　33, 80
Utah, University of　139

Vitalism　127, 149, 164, 197 n. 64
Von Frisch, Karl　21–2, 160, 161,
　　177–8 n. 34

Waist-to-hip ratio (WHR) *see also*
　　beauty, physical　32–3, 72, 73

Wallace, Craig　164
Walters, Sally　35
Ward, Lester　111
Watson, John B.　51, 53, 56, 68, 110,
　　113, 114, 115, 116, 117, 118, 120,
　　126, 127, 128, 132, 148, 149, 164,
　　166
Wellman, Beth　134
Westermarck, Edward　43, 49, 66, 71,
　　72, 86–7, 92–4, 129, 130, 131, 147,
　　150, 176 n. 6
Whiting, Anna Rachael　78, 183–4 n. 82
Whiting, Phineas　184 n. 82
Williams, Carl Easton　67, 72,
　　73, 74
Wilson, Edward O.　1, 2, 13, 21, 23,
　　24–5, 96, 107
Wisconsin, University of　76, 82
Wissler, Clark　53, 54, 120, 123, 125,
　　143
Witty, Paul A.　132
Women, intelligence and
　　173 n. 39
World War I　45, 51–2, 64, 87, 112,
　　118, 120
World War II　43, 46, 146, 155
Wright, Sewall　154

Yale University　12, 52, 55, 57,
　　119, 127
Yanomamö　14
Yerkes Primate Research Center　55,
　　149, 160, 161
Yerkes, Robert Mearns　48–58, 97–8,
　　112, 113, 118–20, 129–30, 142,
　　147, 148, 159, 160, 188 n. 22,
　　191 n. 62

Zahavi, Amotz　24
Zuckerman, Solly　128–9

LaVergne, TN USA
05 April 2010
178173LV00001B/115/P